130/120843 5-4-84

C584

new

£26-45

CW01475315

Photograph of George Huntington at the time he presented his classic paper in 1872. (By courtesy of the National Library of Medicine and Russell N. De Jong, MD, Ann Arbor, Michigan.)

Michael R. Hayden

Huntington's Chorea

Foreword by G. Bruyn

With 69 Figures

Springer-Verlag
Berlin Heidelberg New York 1981

Michael R. Hayden, MB, ChB, PhD, DCH(SA)
Division of Clinical Genetics
Children's Hospital Medical Center
300 Longwood Avenue
Boston, MA.02115, USA

The cover of the book shows two women affected with the dancing mania. It is an engraving by H. Honduis, the younger in 1842 from a drawing by P. Brueghel. The term "chorea" was introduced to medieval literature during the sixteenth century to describe these persons' involuntary movements. The present engraving is no. 3632 Pretenkabinet, Rijksmuseum, Amsterdam, and is reproduced with their permission.

ISBN 3-540-10588-3 Springer-Verlag Berlin Heidelberg New York
ISBN 0-387-10588-3 Springer-Verlag New York Heidelberg Berlin

Library of Congress Cataloging in Publication Data
Hayden, Michael R. Huntington's chorea. Continuation and elaboration of thesis (doctoral) — University of Cape Town, South Africa, 1979. Includes bibliographies and index. 1. Huntington's chorea. I. Title. [DNLM: 1. Huntington chorea. WL 390 H415h] RC394.H85H39 616.8′51′042 81-5712
ISBN 0-387-10588-3   AACR2

© by Springer-Verlag Berlin Heidelberg 1981
Printed in Great Britain

Typeset by Photo-Graphics, Stockland, Honiton, Devon, England
Printed and bound by William Clowes Limited, Beccles and London
2128/3916-54321

For Sandy; and to the memory
of Marjorie Guthrie

Foreword

It is a pleasure to send Dr. Hayden's monograph on its way to as yet unknown but hopefully widespread destinations with all the valedictions a Foreword may contain.

Since I met the author in Cape Town in 1978 I have been struck, on numerous occasions, by the fortuitous combination of an inquisitive mind, a creative drive, a sharp awareness of the historical and social setting of phenomena, and a solid discipline which his personality displays behind a good-natured laugh. If a tree is known by its fruits, both Dr Hayden's PhD thesis and the present monograph afford quite an insight into the *auctor intellectualis.*

The amalgamation of the terrible misery behind scientific facts and the elegantly artistic presentation of this book will leave none of its readers unperturbed. It reminds me of Nietzsche's 'Denn das Schöne ist nichts als des Schrecklichen Anfang, und wir bewundern es so weil es gelassen verschmäht uns zu zerstören' (Beauty is but Horror's beginning, and we admire it because it resignedly spurns to destroy us). The book is a denial, a testimony against Juvenal's spurious 'Stemmata quid faciunt...' (of what value are pedigrees). For it is the very genetical prolongation of misery over the centuries that brought Huntington's chorea to South Africa, Australia and the USA from the shores of sea-faring seventeenth-century England and Holland.

There have appeared some rays of hope, however. Since a small group of people met in Montreal under the auspices of the WFN (whose President was then Dr Macdonald Critchley) to constitute a WFN Research Committee on Huntington's Chorea, research on this disease has intensified considerably. The Research Group has grown to include some forty members from all over the world, who are closely tied by bonds of friendship and common interest. Out of a humble beginning has grown a body with considerable importance, due in no small way to the inspiring activity of one of its members, Mrs. Marjorie Guthrie, who has organised lay organisations in well over ten countries, and whose efforts led to the constitution of a USA Congressional Committee on this disease. Well over 150 articles a year on Huntington's chorea now appear in the scientific journals. Confidence is growing that a solution to the causative riddle of this disease may be expected in the not too distant future. Patients, and their close relatives, in quite a few countries, have come to realise that their active cooperation with scientific efforts will bring that fortunate moment nearer, by the stimulating effect this has on the scientists.

It is this extra dimension which makes the appearance of this monograph so appropriate and should, along with the book's own merits, earn it a wide and warm reception.

Leiden, April 1981   Professor George W. Bruyn, MD, DSc

Preface

Huntington's chorea is a unique progressive neuropsychiatric disease, with protean manifestations. Affected persons may initially seek help from a clincian in one of many branches of medicine including neurology, psychiatry, genetics, internal medicine, paediatrics and family medicine. To all physicians whose task it is to provide continuing care to patients and their families, it remains one of the most difficult, enigmatic and challenging problems in clinical practice.

Huntington's chorea has existed as a separate and well-defined entity for well over 100 years. It is therefore somewhat surprising that no single-author monograph in the English language has ever been entirely devoted to this subject. The need for an up to date text which gives an overview of the state of current knowledge concerning Huntington's chorea is the primary motivation for this book.

In parallel with the surge of interest in the neurosciences in general, there has been a growing awareness of Huntington's chorea and its implications. Despite this, it is still possible for medical students to pass through their entire training without exposure to patients with this disorder. In many instances, the disease has also been neglected in the training of those in neurology, psychiatry and genetics.

The present text is a personal approach to a complex subject. I have chosen to emphasise aspects of the disease which have relevance to the practising clinician. At the same time, I have included details of the history and geographical distribution of Huntington's chorea, which give a broader perspective. Genetic and psychosocial ramifications have been stressed. For these reasons, other mental health professionals, including psychologists and social workers, may find the text useful.

In the course of preparing this monograph I have had to learn a great deal about areas outside my own field of expertise. To the specialist in these areas it may seem that I am stating the obvious; to the generalist it may appear on occasion that I have gone into too much detail. At all times, however, I have kept the clinician in mind and, I hope, given a balanced perspective.

The understanding of neuropsychiatric and genetic diseases is proceeding extremely rapidly. These developments will have major relevance to those managing affected families, and for this reason a chapter on current trends in research is included.

This monograph is a continuation and elaboration of work which formed part of a thesis submitted to the University of Cape Town, South Africa, for which the degree of Doctor of Philosophy in Medicine was awarded in 1979. Over 120 patients with Huntington's chorea have been personally examined and the records of approximately 400 other affected persons carefully scrutinised.

The ultimate goal of research into Huntington's chorea is the elucidation of the primary defect and the development of a cure. More than ever before, there is reason for optimism. In different parts of the world funds are being allocated in increasing amounts for the investigation of this and other genetic disorders. Although still evident, the social stigma of Huntington's chorea is diminishing. Furthermore there is growing awareness that the burgeoning of research activity in this area must be paralleled by improvement in health care facilities for affected families. I am confident that the next decade will witness significant advances in the understanding of Huntington's chorea and its improved management.

Harvard Medical School   Michael R. Hayden
Boston, March 1981

Acknowledgements

This book was begun in Cape Town and finished in Boston. Many people in different parts of the world greatly assisted in its preparation and to all of them I want to express my deep appreciation.

Professor Peter Beighton of the Department of Human Genetics in Cape Town was primarily responsible for my entering this field in the first place. He provided extensive facilities for the preparation of this manuscript, without which it could not have been completed. Numerous other medical colleagues and friends in Cape Town provided continuous support, reviewed various chapters and gave constructive criticism. These include Drs. Jim Macgregor, Molly Nelson, Peter Bundred, Richard Hewlett and my good friends David Goldberg and Philip Melmed.

Dr. David Stevens in England has made a profound contribution to the book. He has shared the results of his own study with me and meticulously reviewed the whole monograph. I have greatly benefited from the enjoyable and spirited discussions I have had with him.

The completion of this monograph has been facilitated by numerous serendipitous circumstances. Amongst the most significant was my first meeting with Professor George Bruyn of Leiden, Holland, in 1978, during which he stressed the need for such a text. His guidance and encouragement since that time have been of inestimable value.

To Professor Richard Goodman of Tel Aviv, who has also given continual support, and Ms. Marjorie Guthrie of New York, who has shown enthusiastic interest in this project, I am deeply grateful.

I am particularly fortunate in having had the patient, good-humoured and untiring secretarial help of Ms. Gillian Shapley. Barbara Breytenbach and Greta Beighton of the Department of Human Genetics in Cape Town have also provided excellent technical and clinical assistance.

I am especially grateful to John Swartz who performed the photography with his usual dedication, expertise and flair, and to Linda Coetzee and Dixon Yun who expertly produced many of the graphics.

Financial support for this project was most generously provided by the Mauerberger Foundation of Cape Town, South Africa.

Numerous colleagues and friends in Boston helped in the final stages of the manuscript. I have very much enjoyed and benefited from talking to Park Gerald, Jack Wood, Drs. James Morris, Ted Bird, John Aldridge, Roy Freeman, Howard Dubowitz and Raun Melmed.

The majority of illustrations are drawn from my own collection, but in certain instances colleagues have kindly provided original prints. I would like to thank the following persons in this regard: Professor R.N. De Jong (Ann Arbor), the Director of the Rijksmuseum, Amsterdam, Professor G. Bruyn (Leiden), Ms. Marjorie Guthrie (New York), Dr. L. Forno (Palo Alto), Drs. J. Morris, T. Bird, F. Gilles and D. Sax (Boston). I am also grateful to the following for permission to use figures: Professor A. Barbeau (Montreal), Dr. P. Harper (Wales), Professor A. Motulsky (Seattle), Dr. E. Spokes (England),

Professor F. Vogel (Heidelberg), the Office of Health Economics in London, Raven Press and Springer-Verlag. The Editors of the *Lancet, Neurology, South African Medical Journal, American Journal of Medical Genetics* and the *Proceedings of the National Academy of Sciences, USA* also granted permission for reproduction of certain illustrations.

Collaborating with Springer-Verlag has been a great pleasure and this book has benefited from Michael Jackson's extensive editorial experience.

My parents and brother have provided outstanding support which is deeply appreciated.

To Sandy, who has been present from the moment of this book's conception, has watched its gestation, and also witnessed and eased it's somewhat painful delivery, I am forever grateful.

Finally to the patients themselves and their families, who have given of their time to share their experience with me, even when they realised that no cure was imminent, I am deeply grateful. Through my interactions with them I have been constantly reminded that the work for this book has not just been a theoretical exercise but rather an enquiry into fundamental aspects of people's lives, their health and illness and their living and dying. If it contributes to better understanding of the disease and thus a possible improvement in the lives of those who are afflicted, its primary aim will have been fulfilled.

Contents

1 Historical Background

The door opened in an irresolute way, and an arm was thrust through with a spasmodic jerk; then a leg followed with a like unsteadiness; while the right hand twitched at the handle the corresponding body clinging to the door edged itself round with a random attempt to close it. Then with a stagger the patient lurched forwards to our table... a tripping, staggering gait, hastening and stopping until he reached the chair. Yet even in the chair, you see as we did, that the man is still in incessant restless motion, like a marionette; now jerking an arm, now the trunk of the body, now shrugging a shoulder and so on.

This vivid description by Allbutt (1918) of the entry of a patient with Huntington's chorea into his consulting rooms graphically portrays the involuntary movements of the disorder. The combination of these abnormal movements, the progressive loss of mental function and the predictable mode of inheritance prompted Davenport and Muncey (1916) to comment that Huntington's chorea is "one of the most dreadful diseases that man is liable to".

Huntington's chorea was definitively described as a separate entity in 1872, but genealogical investigations have established that the disease was present for many years prior to that time. During the preceding centuries there was little differentiation between this and other causes of chorea. For this reason an examination of the historical background of Huntington's chorea would be incomplete without a consideration of the history of chorea as a whole.

The word 'chorea' is derived from the Latin *choreus* alluding to dancing and the Greek *choros* meaning 'chorus'. The description in the Bible of those persons that "reel to and fro, and stagger like a drunken man, and are at their wits' end" (Psalm 107, verse 27) could well fit the clinical picture of a person affected with chorea, even though this was not the intention.

There is evidence in the papyri of ancient Egypt that a loss of voluntary movement was correctly ascribed to brain dysfunction (Edwin Smith papyrus, case 31, 3000 B.C.). However, the realisation that excessive movements could also be due to underlying disease of the central nervous system occurred only 4500 years later. This momentous conceptual leap was made by Paracelsus (1493-1541) in the early sixteenth century and heralded the beginnings of a scientific approach to the investigation of chorea.

The earliest records of involuntary movements pertain to the strange dances that raged through the Middle Ages. Contrary to the popular belief that these dances were due to the Devil's work and only curable by holy men, Paracelsus emphatically stated that the dancing mania was not due to "ghostly beings and spirits" but the "result of some fault in the affected person's personality". This and other revolutionary concepts evoked much abuse and resulted in his colleagues denouncing him as "the Luther of medicine, the vagabond of a pseudo-doctor". He was perturbed by this reaction and in later life was moved to comment that "I've poured out the treasures of my research to those who treat me as a madman and my works as bizarre ravings and I am wounded to the quick" (Thornton 1980).

The superstitions relating to involuntary movements which began with the dancing mania did not die with Paracelsus, and persisted through the centuries that followed. In some instances these misguided beliefs have remained, albeit in a modified form, to the present time. An account of the history of involuntary movements is in Barbeau's words (1958) "not only a story of man's credulity and ignorance, but also concerns one of the most difficult problems confronting neurologists and physiologists. For the most part, it is not the story of great men, but that of astute observers, simple and honest practitioners devoted to the care of sufferers."

1.1. Chorea in the Middle Ages: The Dancing Mania

The Middle Ages witnessed one of the most destructive epidemics in our history, namely the plague which devastated a quarter of the population of Europe and thus earned the ignominious title of the Black Death. This event formed the background to the beginning of a "strange new epidemic—the dancing mania", which began in 1374 in Aix-la-Chapelle (Aachen) on the Franco-German border and which within a few months had spread to Cologne and to the neighbouring Netherlands, France and Belgium.

Hecker's (1888) historical review of 'the dancing mania' is still the standard accepted authority on this subject. He vividly described how victims "formed circles hand-in-hand,

Fig.1.1. Sixteenth-century drawing depicting "The pilgrims who danse on St.John's day in Meulenbeeck, outside Brussels. If they danse or jump over a bridge they are cured for a year from St.John's illness." Drawing after Pieter Brueghel I. (No. 16859, by courtesy of the Rujksmuseum, Amsterdam)

appearing to have lost all control over their senses and continued dancing regardless of the bystanders for hours together in wild delirium, until at length they fell to the ground in a state of exhaustion... While dancing they neither saw nor heard but were haunted by visions, their fancies conjuring up spirits whose names they shrieked out."

In Germany this phenomenon was first called St. Johannes (John's) chorea. St. John the Baptist was the patron saint who guarded against epilepsy and other movement disorders, and the epidemic of dancing that gripped many persons in the late fourteenth century was given his name in the hope that this might cure them of the disorder. St. John's day was customarily celebrated with different rituals, such as leaping through a fire or dancing over a particular bridge (Fig. 1.1), in the belief that the participants would thereby be protected from harmful illnesses for the coming year.

St. John's chorea spread rapidly and reached Strasburg around 1418, after which the name of St. Vitus chorea and its synonyms, morbus St. Viti and chorée de St. Guy, became attached to the disorder (Lewis-Jonsson 1949). The personal history of St. Vitus is crucial to the understanding of this eponymous designation for the dancing mania. According to legend St. Vitus was a Sicilian youth who died as a martyr in southern Italy during the persecutions of the Christians in the fourth century. Shortly before being put to death in a cauldron of boiling lead and pitch, Vitus is said to have prayed to God to save all those from the dancing mania who fasted for the evening prior to his day (15 June). He is supposed to have immediately heard a voice from heaven saying "Vitus, thy prayer is accepted" (Bett 1932). In this way Vitus became the patron saint of all afflicted with the 'dancing plague'.

The 'disease' spread to Italy, where it was known as tarantism as it was supposed to have been caused by the bite of the tarantula. It is interesting that a similar disorder which mimicked the dancing mania was described centuries later in Ethiopia, occurring particularly in the province of Tigre and thus known as Tigretier. More recent equivalents of the dancing mania are to be found in the description of the epidemic of chorea in the neighbourhood of Maryville, Tennessee, in 1805 (Robertson) and the account of the 'leaping ague' in localised areas of Scotland in 1807 (The Inquirer).

It is generally accepted that the dancing mania of the Middle Ages did not have an organic basis, but was rather a manifestation of hysteria. The frequency of St. Vitus' chorea or tarantism decreased in the sixteenth century as the circumstances which precipitated the disorder improved and changed. However, it is noteworthy that there is, to this day, an annual procession of dancers in Meulenbeeck on St. John's Day (L. Went, 1978, personal communication).

The term 'chorea' was introduced to medical literature by Paracelsus in the sixteenth century. He investigated abnormal movements more closely and subdivided chorea into three types, namely: chorea imaginativa aestimative, which arises from the imagination and is thought to represent the dancing mania; chorea lasciva, which results from 'sensual desires'; and chorea naturalis coatta, which has an organic basis. Bell (1934) has suggested that some patients in this last group may have had Huntington's chorea.

The next major contribution to the classification and delineation of chorea was made 150 years after Paracelsus by Thomas Sydenham.

1.2. Thomas Sydenham and Chorea

Thomas Sydenham, an Englishman of noble descent, was born in the village of Wynford Eagle in 1624 and is generally recognised as the father of chorea. He obtained his licence to practise medicine in 1663, when he was 39 years of age, and despite this late start made

Thomæ Sydenham, M.D.

OPERA

UNIVERSA.

In quibus non folummodò Morborum
Acutorum Hiftoriæ & Curationes novâ & ex-
quifitâ methodo diligentiffimè traduntur, ve-
rùm etiam Morborum ferè omnium Chronico-
rum Curatio breviffima, pariter ac fideliffima
in Publici commodum exhibetur.

---

*Editio altera, priori multùm auctior, &
emendatior reddita.*

Huic etiam de novo acceffit Index Alphabe-
ticus fummam omnium rerum, & Curatio-
num fingularum, in gratiam ftudioforum,
breviter complectens.

LONDINI,

Typis *R.N.* impenfis *Walteri Kettilby,* ad Infigne
Capitis Epifcopalis in Cœmeterio D. *Pauli,* 1685.

Fig.1.2. Title page of the first collected edition of Sydenham's *Works,* 1685

an enormous contribution to modern medicine through his astute observations and excellent clinical judgement (Fig. 1.2). These abilities eventually earned him the title of the 'English Hippocrates'.

His excellent, concise description of chorea in 1686 resulted in the symptom-complex bearing his name and began as follows:

> St. Vitus' dance is a sort of convulsion which attacks boys and girls from the tenth year until they have done growing. At first it shows itself by a halting or rather an unsteady movement of one of the legs, which the patient drags. Then it is seen in the hand of the same side. The patient cannot keep it a moment in its place, whether he lay it upon his breast or any part of his body. Do what he may it will be jerked elsewhere convulsively.

He concluded by suggesting that "this affection arises from some humour falling on the nerves and such irritation causes the spasm".

While his description represented a great advance in the understanding of chorea, by erroneously adopting the name of St. Vitus' dance, which referred to the dancing mania, he added to the existing nosological confusion. This error has been condoned by nearly 300 years of continual usage. The movements described by Sydenham became known as 'Sydenham's chorea', 'chorea anglorum' or 'chorea minor' (Fig. 1.3). The name chorea anglorum was given in honour of the outstanding contributions made by English physicians to the understanding of chorea. The term chorea minor was used to denote the above-mentioned disorder and differentiated this entity from 'chorea major', which referred

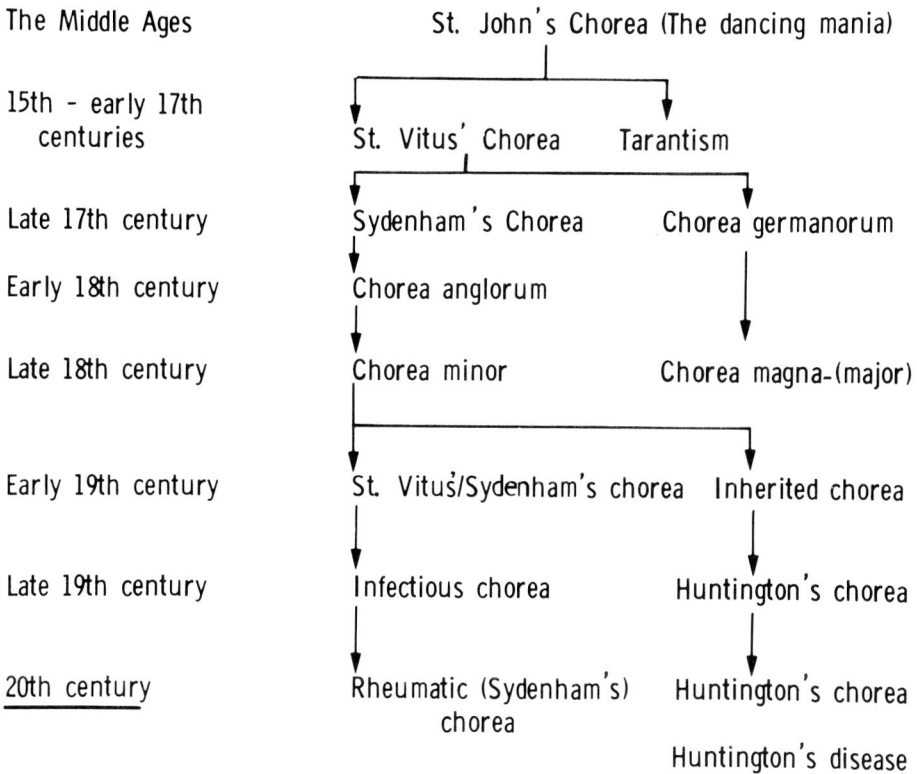

The Middle Ages	St. John's Chorea (The dancing mania)	
15th – early 17th centuries	St. Vitus' Chorea	Tarantism
Late 17th century	Sydenham's Chorea	Chorea germanorum
Early 18th century	Chorea anglorum	
Late 18th century	Chorea minor	Chorea magna-(major)
Early 19th century	St. Vitus/Sydenham's chorea	Inherited chorea
Late 19th century	Infectious chorea	Huntington's chorea
20th century	Rheumatic (Sydenham's) chorea	Huntington's chorea Huntington's disease

Fig.1.3. The historical development of the nosology of chorea

to the dancing mania. The latter affliction was also recognised by the eponym 'chorea germanorum' in view of its higher frequency in Germany compared with other European countries.

Most of the publications concerning chorea during the seventeenth century were either in German or English and it was not until 1810 that Bouteille in his *Traité de la chorée, ou danse de St. Guy* brought attention to this subject in France and also reclassified chorea as follows: chorée essentielle, affecting children between the ages of 10 and 14 years; chorée secondaire, occurring at any age, secondary to some other disease; and chorée fausse, which only resembles chorea, but is not a true example of this symptom.

Despite these attempts at nosological clarity, confusion concerning the nomenclature of abnormal movements still prevailed in the nineteenth century, as evidenced by the 15 synonyms for chorea listed in Dunglison's textbook of medicine of 1842. William Osler (1894) expressed his dismay in the opening lines of his book *On Chorea and Choreiform Affections* as follows: "In the whole range of medical terminology there is no such olla podrida as chorea, which for a century has served as a sort of nosological pot into which authors have cast indiscriminately." This confused situation was dramatically improved once it became generally accepted that certain forms of chorea were inherited.

1.3. From Sydenham to Huntington: The First Descriptions of Inherited Chorea

The first definite mention of heredity in the causation of adult chorea was made by Elliotson, an English physician in 1832 (Stevens 1972). In a lecture on St. Vitus' dance published in the *Lancet,* he noted that "when it [chorea] occurs in adults it is frequently connected with paralysis or idiotism and will perhaps never be cured. It appears to arise for the most part from something in the original constitution of the body, for I have often seen it hereditary." It is most likely that Elliotson was describing what is known today as Huntington's chorea. The suggestion that chorea could be inherited was made some years prior to Elliotson's contribution (Armstrong 1783; Bernt 1810; Mongenot 1815) but in these instances chorea in childhood or Sydenham's chorea was the more likely diagnosis.

The Rev. Dr. Charles Oscar Waters on graduation in 1841 sent a letter from Franklin, New York, to Dr. Robley Dunglison, his Professor of Medicine at Jefferson Medical College, describing a disorder "which is markedly hereditary... very rarely makes its appearance before adult life... in all cases induces a state of more or less perfect dementia... and never ceases while life lasts". Waters mentioned that the disease was known among the common people as the 'magrums', which was probably derived from the word 'megrim' and referred to the staggering and fidgeting of affected individuals. Dunglison published Waters' letter in full in the first edition of his textbook in 1842.

The next person to describe this disease was Charles Gorman of Philadelphia as part of his doctoral dissertation in 1846, which was presented before the Faculty of Jefferson Medical College and entitled 'On a Form of Chorea, Vulgarly Called Magrums'. The original thesis was lost during the demolition of the Old Jefferson Colleges and thus the details of his account will probably never be known.

Another important episode in the early history of inherited chorea relates to a district physician in the Saetersdal District of Norway, Johan Christian Lund, who gave a very accurate description of the disease in 1860. The account was in Norwegian and was part of his report on health and medical conditions in Norway. It is a matter of regret that, probably due to language barriers, this contribution remained largely unrecognised for almost 100 years, until Ørbeck translated it into English in 1959. Ørbeck noted that in view of the accuracy of Lund's description "it is rather surprising that Lund's name is not mentioned in monographs on Huntington's chorea". Another possible factor in the failure of this article to excite attention was that Hirsch, author of the famous and influential *Handbook of Historical and Geographic Pathology* (1860) incorrectly described the disorder reported by Lund as corresponding to paralysis agitans. In Norway today, however, Huntington's chorea is still sometimes referred to as Lund-Huntington's chorea.

The next mention of the disorder was in an article entitled 'Chronic Hereditary Chorea', written in 1863 by Lyon during the time he was house physician in New York City. He gave details of three affected families, and the disease described by Lyon is certainly the same as that documented previously by Elliotson, Waters, Gorman and Lund and would be recognised today as Huntington's chorea. In spite of these earlier descriptions the inherited form of chronic chorea was not generally recognised until after Huntington's contribution.

1.4. Huntington's Chorea

George Huntington was born in East Hampton, Long Island, New York, on 9 April 1850 (Fig. 1.4). Both his father and his grandfather were medical doctors and Huntington himself reported (1910) that his grandfather had already observed patients with inherited chorea when he moved to East Hampton in 1797. George Huntington was only 8 when he saw his first cases of the disorder, whilst riding with his father on his professional rounds. Later, in 1909, he described this experience to the New York Neurological Society as follows:

> I recall it as vividly as though it had occurred but yesterday. It made a most enduring impression upon my boyish mind which was my very first impulse to my choosing chorea as my virgin contribution to medical lore. Driving with my father through a wooded road leading from East Hampton to Amagansett we suddenly came upon two women, mother and daughter, both tall, thin, almost cadaverous, both bowing, twisting, grimacing. I stared in wonderment, almost in fear. What could it mean? My father paused to speak with them and we passed on. Then my Gamaliel-like instruction began; my medical education had its inception. From this point on, my interest in the disease has never wholly ceased.

This early exposure to the disorder was probably one of the major events which encouraged him to study medicine, and only a year after graduating from Columbia University at the age of 21, he presented his paper 'On Chorea' to the Meigs and Mason Academy of Medicine in Middleport, Ohio. The text of the lecture appeared in the *Medical and Surgical Reporter* of Philadelphia on 13 April 1872 (Fig. 1.5). In view of his considerable expertise and penetrating insight, it is remarkable that this was his sole contribution to medical literature. After a general discussion on the subject of chorea he described the hereditary form of the disorder. On account of its brevity and excellence, extracts from Huntington's paper are cited here:

> And now I wish to draw your attention more particularly to a form of the disease which exists, as far as I know, almost exclusively in the east end of Long Island... The hereditary chorea, as I shall call it, is confined to certain and fortunately few families and has been transmitted to them, an heirloom from generations way back in the dim past. It is spoken of by those in whose veins the seeds of the disease are known to exist, with a kind of horror, and not at all alluded to except through dire necessity, when it is mentioned as 'that disorder'. It is attended generally by all the symptoms of common chorea, only in an aggravated degree, hardly ever manifesting itself until adult or middle life, and then coming on gradually but surely, increasing by degrees and often occupying years in its development, until the hapless sufferer is but a quivering wreck of his former self.
>
> It is as common and is, indeed, I believe, more common among men than women, while I am not aware that season or complexion has any influence in the matter. There are three marked peculiarities in this disease: 1. its hereditary nature; 2. a tendency to insanity and suicide; 3. its manifesting itself as a grave disease only in adult life.

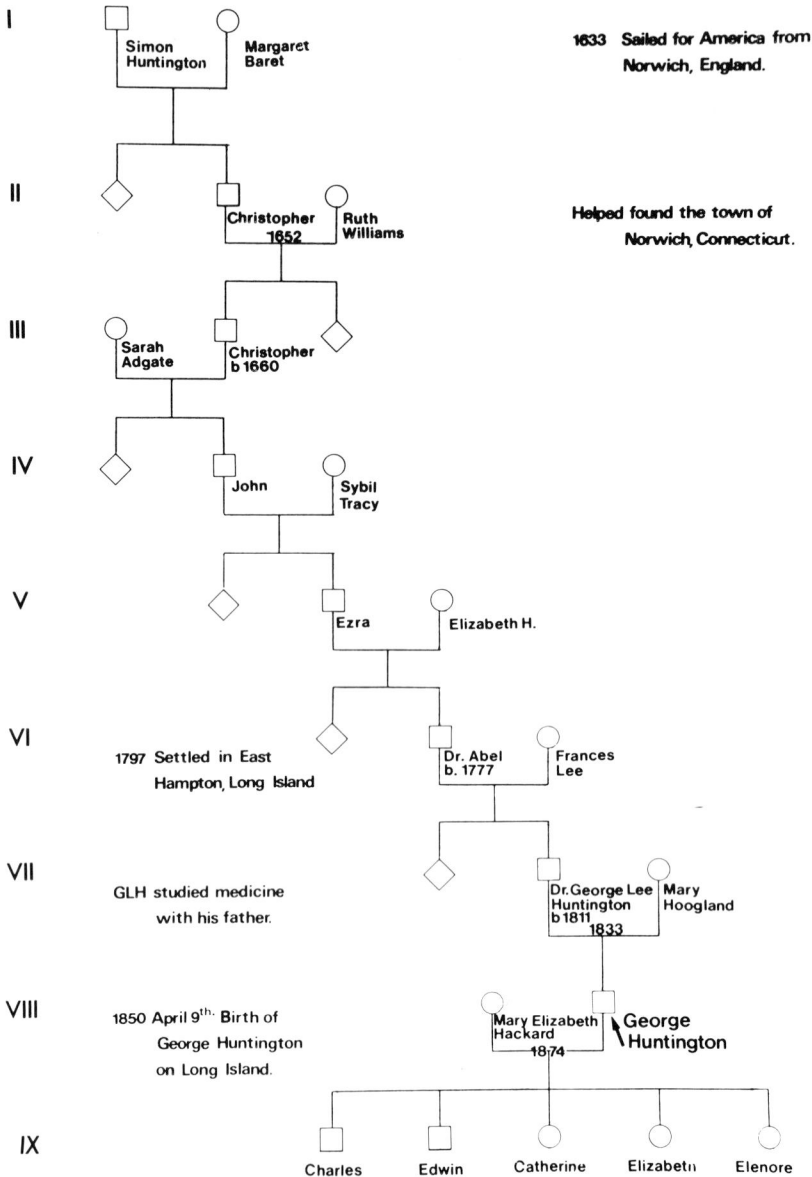

Fig. 1.4. The ancestral tree of George Huntington

He continued his paper by discussing, with examples from his experience, each of these three 'peculiarities'. He accurately defined the inheritance of the disorder, which was consistent with an autosomal dominant trait. The antisocial behaviour and lack of insight of affected persons were vividly portrayed in his description of "two married men, whose wives were living and who were constantly making love to some young lady, not seeming to be aware of any impropriety in it. They are men of about 50 years of age, but never let

MEDICAL AND SURGICAL REPORTER.

No. 789.]   PHILADELPHIA, APRIL 13, 1872.   [Vol. XXVI.—No. 15.

ORIGINAL DEPARTMENT.

Communications.

ON CHOREA.

By George Huntington, M. D.,
Of Pomeroy, Ohio.

Essay read before the Meigs and Mason Academy of Medicine at Middleport, Ohio, February 15, 1872

Chorea is essentially a disease of the nervous system. The name "chorea" is given to the disease on account of the *dancing* propensities of those who are affected by it, and it is a very appropriate designation. The disease, as it is commonly seen, is by no means a dangerous or serious affection, however distressing it may be to the one suffering from it, or to his friends. Its most marked and characteristic feature is a clonic spasm affecting the voluntary muscles. There is no loss of sense or of volition attending these contractions, as there is in epilepsy; the will is there, but its power to perform is deficient, the desired movements are after a manner performed, but there seems to exist some hidden power, something that is playing tricks, as it were, upon the will, and in a measure thwarting and perverting its designs; and after the will has ceased to exert its power in any given direction, taking things into its own hands, and keeping the poor victim in a continual jigger as long as he remains awake, generally, though not always, granting a respite during sleep. The disease commonly begins by slight twitchings in the muscles of the face, which gradually increase in violence and variety. The eyelids are kept winking, the brows are corrugated, and then elevated, the nose is screwed first to the one side and then to the other, and the mouth is drawn in various directions, giving the patient the most ludicrous appearance imaginable.

The upper extremities may be the first affected, or both simultaneously. All the voluntary muscles are liable to be affected, those of the face rarely being exempted.

If the patient attempt to protrude the tongue it is accomplished with a great deal of difficulty and uncertainty. The hands are kept rolling—first the palms upward, and then the backs. The shoulders are shrugged, and the feet and legs kept in perpetual motion; the toes are turned in, and then everted; one foot is thrown across the other, and then suddenly withdrawn, and, in short, every conceivable attitude and expression is assumed, and so varied and irregular are the motions gone through with, that a complete description of them would be impossible. Sometimes the muscles of the lower extremities are not affected, and I believe they never are *alone* involved. In cases of death from chorea, all the muscles of the body seem to have been affected, and the time required for recovery and degree of success in treatment seem to depend greatly upon the amount of muscular involvement. Romberg refers to two cases in which the muscles of *respiration were affected*.

The disease is generally confined to childhood, being most frequent between the ages of eight and fourteen years, and occurring oftener in girls than in boys. Dufosse and Rufz refer to 420 cases; 130 occurring in boys and 299 in girls. Watson mentions a collection of 1,029 cases, of whom 733 were *females*, giving a proportion of nearly 5 to 2. Dr. Watson also remarks upon the disease being most frequent among children of *dark* complexion, while the two authorities just alluded to, Dufosse and Rufz, give as their opinion that it is most frequent in children of *light* hair. In every case visiting the clinics

Fig. 1.5. The first page of Huntington's article 'On Chorea', published in 1872

an opportunity to flirt with a girl go past unimproved. The effect is ridiculous in the extreme."

His account of the disorder was concise and accurate and had the good fortune of being immediately abstracted into German (Kussmaul and Nothnagel 1872), with the result that Huntington's name very quickly became attached to the disease in different parts of the world, including Germany (Huber 1887), France (Klippel and Ducellier 1888), Italy

Table 1.1. Landmarks in the history of Huntington's chorea

1832	John Elliotson, the English physician, first mentioned the possible influence of heredity on adult chorea
1841	The Rev. Dr. Waters (1816-1892) described adults with hereditary chorea in the New York area in a letter to his Professor of Medicine
1842	Charles Gorman submitted a dissertation 'On a Form of Chorea Vulgarly Called Magrums'
1858	George Huntington's first exposure to affected patients whilst accompanying his father on his rounds
1860	Johan Christian Lund, district physician in Norway, described chorea St. Vitus which "recurs as a hereditary disease in Saetersdal"
1863	Irving Lyon submitted article on chorea to *American Medical Times*, describing affected families in Connecticut
1872	George Huntington's definitive description of the disease
	George Huntington's article abstracted into German
1887	Huber first used eponym Huntington's chorea to describe hereditary chorea
1888-1895	Eponym adopted by different authors in France, Italy, England and USA
1908	First monograph on Huntington's chorea (edited by Browning), citing over 200 references
1909	George Huntington addresses the New York Neurological Society
1916	The death of George Huntington
1967	The first meeting of the World Federation of Neurology Subgroup on Huntington's Chorea in Montreal, Canada. Meetings subsequently held biennially
1972	Centennial Symposium on Huntington's chorea in Columbus, Ohio, USA, in commemoration of Huntington's original contribution
1977	The Commission for the Control of Huntington's Disease in the USA reports to Congress and the President
1978	Second International Symposium on Huntington's Disease held in San Diego, California, USA

(Seppili 1888) and England (Suckling 1889). Sir William Osler was moved to comment in 1908 that "there are few instances in the history of medicine in which a disease has been more accurately, more graphically or more briefly described".

In spite of these accolades, Huntington did not follow an academic career. At the time he certainly did not realise the significance of his contribution and mentioned to the Meigs and Mason Academy that "I have drawn your attention to this form of chorea, gentlemen, not that I considered it of any great practical importance to you, but merely as a medical curiosity and as such it may have some interest".

By all accounts he was a modest, sensitive, artistic man with a well-developed sense of humour. He spent much of his free time in the woods, sketching the animals (Maltsberger 1961). He was also a keen amateur musician and often played the flute to his wife and five children. To his patients he was kind, a virtue epitomised in the story of his dealings with a child of an impoverished family, who died of diphtheria. The father was ill and the mother overcome with grief, so Huntington himself built the coffin and organised the burial (Stevenson 1934).

It is noteworthy that the pencilled remarks of Huntington's father are still visible in the margins of the original manuscript (De Jong 1973) and it is clear that the eventual delineation of the disorder was the outcome of the cooperation between the three generations of doctors in the Huntington family. It is fitting that the name 'Huntington' should have gained acceptance as the eponymic designation of the disorder.

In this context it is relevant to mention here that there is a trend to move away from eponyms in the description of different disorders. However, in certain instances the eponym has been retained as it has been hallowed by time and received universal acceptance. In these circumstances it is no longer customary to use the possessive form of the person's name, e.g. Huntington's chorea, as the person in question neither suffers from the disease nor owns it. However, for the sake of convenience and clarity the term 'Huntington's chorea' has been retained in this monograph. Also, in recent times the name 'Huntington's disease' has gained more widespread acceptance, particularly in the USA. This change in nomenclature has been influenced by the increasing awareness that rigidity is an important feature of the disorder, as a result of which it has been suggested that the name Huntington's chorea is too limited and causes confusion. However, there is at present no international consensus concerning the correct terminology of this disease and as the term Huntington's chorea is still currently in usage in most parts of the world it is used here.

Four of the first six accounts of adults with inherited chorea emanated from America (Table 1.1). The early and sustained interest in Huntington's chorea in that country resulted in more frequent identification of affected persons and led some workers to believe that the disorder was primarily an "American tragedy, more sinister and far more interesting than anything imagined by Theodore Dreiser" (Critchley 1934). However, this situation reflected greater awareness of the disorder in America rather than true geographical differences in prevalence.

References

Allbutt C (1918) Notes from a clinical lecture on a case of Huntington's chorea. Br Med J i: 389–390

Armstrong F (1783) Account of singular convulsive fits in three children of one family. Medical Commentaries 9: 317–325

Barbeau A (1958) The understanding of involuntary movements: an historical approach. J Nerv Ment Dis 127: 469–489

Bell J (1934) Huntington's chorea. In: Fisher R A (ed) The treasury of human inheritance, vol IV/1. Cambridge University Press, London, pp 1–77

Bernt J (1810) Monographia choreae Sancti Viti. Prague

Bett WR (1932) Some pediatric eponyms. IV. Sydenham's chorea. Br J Child Dis 29: 283

Bouteille GM (1810) Traité de la chorée ou danse de St. Guy. JB Ballière, Paris, pp 1–362

Critchley M (1934) Huntington's chorea and East Anglia. J State Med 42: 575–587

Davenport CB, Muncey EB (1916) Huntington's chorea in relation to heredity and eugenics. Am J Insan 73: 195–222

De Jong RN (1953) George Sumner Huntington (1850–1916). In: Haymaker W (ed) Founders of neurology. Charles C Thomas, Illinois, pp 305–307

De Jong RN (1973) The history of Huntington's chorea in the United States of America. In: Barbeau A, Chase TN, Paulson GW (eds) Advances in neurology, vol 1. Raven Press, New York, pp 19–27

Dunglison R (1842) The practice of medicine, 1st edn. Lee and Blanchard, Philadelphia, pp 1–312

Elliotson J (1832) St Vitus's dance. Lancet I: 162–165

Gorman CR (1846) On a form of chorea, vulgarly called magrums. Thesis, Jefferson Medical College, Philadelphia

Hamilton AS (1908) A report of 27 cases of chronic progressive chorea. Am J Insan 64: 403–475

Hecker JFC (1888) The black death. In: Morley H (ed) The epidemics of the Middle Ages. Cassell, London, pp 105–191

Hirsch A (1860) Handbuch der Historisch-geographischen Pathologie, B III. Ferdinand Enke, Erlangen, p 572 (Quoted by Ørbeck)

Huber A (1887) Chorea hereditaria der Erwachsenen (Huntington's chorea). Virchows Arch [Pathol Anat] 108: 267–285

Huntington G (1872) On chorea. Med Surg Rep 26: 317–321

Huntington G (1910) Recollections of Huntington's chorea as I saw it at East Hampton, Long Island, during my boyhood. NY Neurological Society, Dec 1909 (Also in J Nerv Ment Dis 37: 255)

The Inquirer (1807) Edin Med J 3: 434-447

Klippel M, Ducellier F (1888) Un cas de chorée hereditaire de l'adulte (maladie de Huntington). Encephale 8: 716-723

Kussmaul A, Nothnagel CWH (1872) In: Virchow-Hirsch's Jahrbuch für 1872. Berlin, p 175

Lewis-Jonsson J (1949) Chorea: its nomenclature, etiology and epidemiology in a clinical material from Malmohus County (1910-1944). Acta Paediatr [Suppl] 16: 1-140

Lund JC (1860) Chorea Sancti Vidi i Saeterdalen uddrag af distrikslaege for 1860. Gorholdene Norge 1860. Norges Off. Statist. C, 4: 137-138. Aaret 1862

Lyon JW (1863) Chronic hereditary chorea. Am Med Times 7: 289-290

Maltsberger JT (1961) Even unto the 12th generation: Huntington's chorea. J Hist Med 16: 1-17

Mongenot R (1815) See Lond Med Phys J (1820) 44: 68

Ørbeck AL (1959) An early description of Huntington's chorea. Med Hist 3: 165-168

Osler W (1894) On chorea and choreiform affections. Blakistan, Philadelphia, pp 1-175

Osler W (1908) Historical note on hereditary chorea. Neurographs 1: 113-116

Robertson F (1805) An inaugural essay on Chorea Sancti-Viti. Philadelphia (Quoted by The Inquirer, 1807)

Seppili G (1888) Corea ereditaria (Corea d'Huntington—corea cronica progressiva). Riv Sper Freniat 13: 453-459

Stevens DL (1972) The history of Huntington's chorea. J R Coll Physicians Lond 271-282

Stevenson CS (1934) A biography of George Huntington, MD. Bull Inst Hist Med 2: 53-76

Suckling CW (1889) Hereditary chorea (Huntington's disease). Br Med J ii: 1039

Sydenham T (1848-1850) The entire works of Thomas Sydenham. Sydenham Society, London, p 2

Thornton NT (1980) The power of Paracelsus. Hist Med 8: 10-15

Von Hohenheim, PATPB (1658) Opera omnia medico-chimico-chirurgia, edn novissima. J Antonii BS de Tournes, Genoa, pp 1-110

Waters CO (1842) Letter dated 5 May 1841. In: Dunglison R (ed) Practice of medicine, 1st edn. Lee and Blanchard, Philadelphia, p 245

2 Genealogy and Geographic Distribution

Genealogy is the study of family connections and as such has major relevance to all genetic diseases. Huntington's chorea, in particular, offers fertile ground for genealogical enquiry because of its distinctive characteristics, its long history and the fact that the mutation rate is amongst the lowest recorded for any inherited disease (Sect. 7.1).

The word 'genealogy' is derived from two Greek words: *genos* meaning lineage or family and *logos* meaning knowledge. The literal meaning of genealogy is the acquisition of knowledge of a lineage. In the narrow sense this process comprises an investigation of a person's ancestors, but in its broadest perspective it is a study of a person and his family in their particular social and cultural environment.

2.1. Genealogical Methods

The most important aid to establishing the diagnosis of Huntington's chorea is confirmation of a possible family history of the disorder. In many instances this will be self-evident, but in others it may prove extremely difficult. The first step in the collection of genealogical information is the family interview. Whilst the principles of obtaining a family history are similar with regard to all genetic diseases, Huntington's chorea poses specific problems. In view of its tragic and far-reaching social effects, denial of the existence of the disease, even when relatives have been admitted to mental hospitals, is common. Family members are often only willing to reveal their ancestry once a relationship of trust and friendship develops.

The construction of a pedigree customarily begins by identifying the proband (the affected person first brought to notice) and then proceeds logically to the sibs and parents. The standard symbols shown in Fig. 2.1 can be used to record the information obtained. Where possible the full name and address, maiden name where applicable, birth date, birth place and date of death are noted. Specific questions with regard to suicide, a full reproductive history and evidence of antisocial behaviour are important and should be approached with care. The scope of the enquiry can be broadened to include details of the natural history and clinical features of the disease. Family bibles, old letters, diaries, birthday books and gravestones can then be examined in an effort to trace further the kinship's history. Photographs and old family albums are also very useful sources of information (Fig. 2.2). Examination of sequential photographs may reveal important details concerning the course of the disease in affected persons.

It may be most helpful to spend time with the oldest living member of the family. These conversations can be rewarding not only for the fascinating insights they give into a way of life long past, but also because they may provide medically important details.

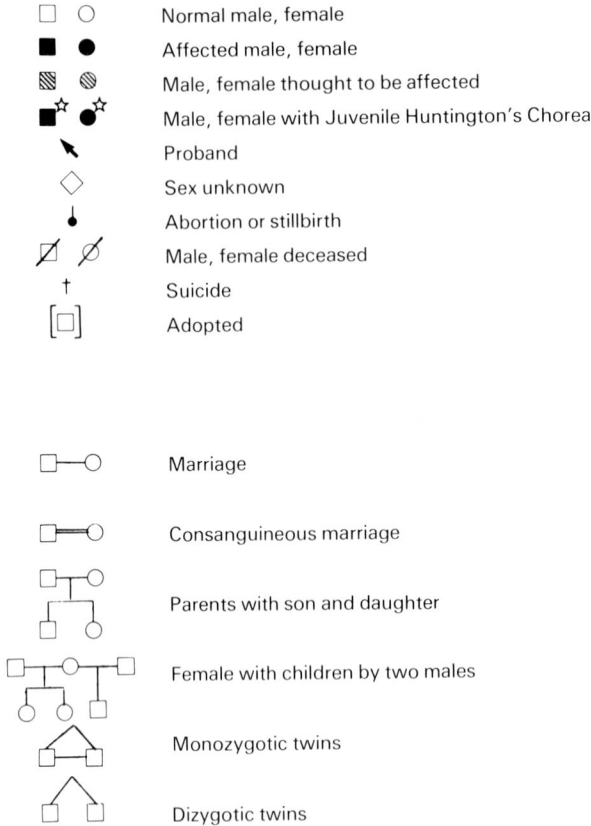

□ ○   Normal male, female
■ ●   Affected male, female
▨ ◍   Male, female thought to be affected
■☆ ●☆   Male, female with Juvenile Huntington's Chorea
↖   Proband
◇   Sex unknown
•   Abortion or stillbirth
⌀ ⌀   Male, female deceased
†   Suicide
[□]   Adopted

□—○   Marriage

□═○   Consanguineous marriage

□┬○   Parents with son and daughter
□ ○

□┬○┬□   Female with children by two males
○ ○ □

△   Monozygotic twins

⟨   Dizygotic twins

Fig. 2.1. Symbols commonly used in the drawing of pedigrees

Clues as to the presence of the disorder in different family members may be expressed anecdotally with such comments as 'Uncle Peter's sudden change in mood' or how 'Grandfather James walked like a drunken man, but we all knew he never drank'.

As a result of these interviews and historical documents it may be possible to unravel the history of a particular kinship and understand some of the factors which have influenced these people's major decisions, such as curtailment of reproduction or emigration. Furthermore, persons previously thought to be unrelated can sometimes be shown to be linked in previous generations, through a connection which often spans oceans and national boundaries.

After the relevant information has been gleaned from the family interviews, the government and church archives can be consulted in an effort to trace the lineage. In some countries, such as the USA and the UK, there are comprehensive genealogical records which will considerably aid such an investigation. Where these documents are incomplete or unavailable, reliance will have to be placed more heavily on the family consultation.

There will be instances when apparently healthy parents have produced a child affected with Huntington's chorea. In most cases this represents unreliable information about the health of the parents or their early accidental death before the disease became manifest. Before suspecting a new mutant, it is wise to remember Bruyn's (1968) advice "to search laterally, search the ascendants and the descendants" in an effort to find hidden clues to the possible presence of the disorder in other family members. Finally, the aphorism

Fig.2.2. This photograph from a family album shows a gentleman with a leaning posture, a newly grown beard, and a characteristic positioning of his right hand, which might be evidence for the presence of Huntington's chorea

'pater semper incertum est' alerts one to the possibility that the patient may not be the offspring of both alleged parents, but rather the illegitimate child of an unsuspected and unknown father who had the abnormal gene for Huntington's chorea.

Whilst this may be true in some instances, it must be realised that a small proportion of cases do, in fact, arise via mutation. Stevens (1976) has estimated that mutation accounted for around 1% of all affected persons in his study. If one extrapolates from his figures for Leeds, England, there are between 15 and 250 affected persons in Britain today who acquire the gene in this way.

A major problem confronting the genealogist in Huntington's chorea is the failure of adequate documentation of this disorder before 1872. As a result, clues as to the prior presence of Huntington's chorea have to be sought by indirect means, including the person's behaviour, personality, habits and ability to perform different tasks. A sudden alteration in personality, with a history of antisocial behaviour, promiscuity, a sudden inability to do the manual labour he was used to, and a change in handwriting, may be signs that the disease is indeed present. The tracing of the origins of the gene for Huntington's chorea is thus time-consuming and demands meticulous, patient attention to all details, however irrelevant they might seem at the time.

2.2. Genealogical Investigations in Various Parts of the World

In this section the classic genealogical investigations concerning Huntington's chorea are reviewed. These studies have taken place in several regions including North America, Europe, Australia and South Africa. The methodology for these projects has differed according to the facilities available in each country. Even though it may be possible to identify the earliest transmitter of the gene for Huntington's chorea in different countries, it is certain that there have been other sources of the abnormal gene in those areas. These include those instances of Huntington's chorea arising via mutation and those affected persons who can be traced to other immigrants.

2.2.1. United States of America

The earliest and most detailed genealogical studies of Huntington's chorea emanated from the USA, and in many instances these have served as models for similar investigations around the world.

As a result of patient genealogical research Jelliffe (1908) was able to link up a considerable number of scattered and apparently unrelated patients who had the disorder. He established that there was a minimum of three separate groups of patients to which most affected persons on the eastern seaboard of the United States could be traced. They are listed as follows:

Group I.   Those in the East Hampton area, which included George Huntington's original patients.

Group II.   The Bedford or Westchester County group, which included those persons reported by Waters and Lyon.

Group III. Those persons who came from Long Island and included the patients described by Gorman. These became known as the Wyoming group.

Tilney (1908) studied affected kindred from the Bedford group and traced their lineage over 200 years to a family who had landed in colonial Connecticut in 1635. Hattie (1909) reported a kinship with several affected persons who had fled from Montbéliard near the Franco-Swiss frontier when the Edict of Nantes was revoked in 1685. After a short stay in South Carolina, the descendants settled in the village of Tatamagouche, Nova Scotia, in 1771. Dynan (1914) traced most of his 19 patients to Germany, though some had emigrated from the United Kingdom, France and Holland. Davenport and Muncey (1916) made an extensive study of 962 persons with Huntington's chorea and traced many patients to four distinct early-immigrant families who arrived in the Salem and Boston areas from Europe in the early part of the seventeenth century. He showed that the progenitors of these people emigrated westwards from the eastern centres as far as Nebraska, Oregon and California. Migration is continuing and Critchley (1973) has noted that a family group living in Hawaii is related to a person who left the Pacific coast in 1824 and whose ancestors had previously lived in New England.

Vessie (1932) studied the colonial records in the public archives of Connecticut and Massachusetts and identified the earliest affected persons in America as three men to whom he gave the pseudonyms Jeffers, Nicolas and Wilkie. These men originally had lived in the village of Bures in East Anglia, England (Fig. 2.3). They were members of the John Winthrop fleet which set sail in 11 ships from Yarmouth and arrived in Salem 3 months later in 1630. The tragic events surrounding their lives in England, their voyage and early days in America are interwoven with the fundamental beliefs of those times and for this reason warrant further consideration.

The first decades of the seveenth century were times of great upheaval in England. The reigning monarch, King James I, was a good-natured, genial man who had, however, little tolerance of nonconformity. When approached by a group of petitioners in 1604 he cried "I shall make them conform or I will harry them out of the land" (Trevelyan 1945).

Superstitions regarding witchcraft exerted much influence at the time. King James I was obsessed with eradicating those thought to be witches, on the grounds that such

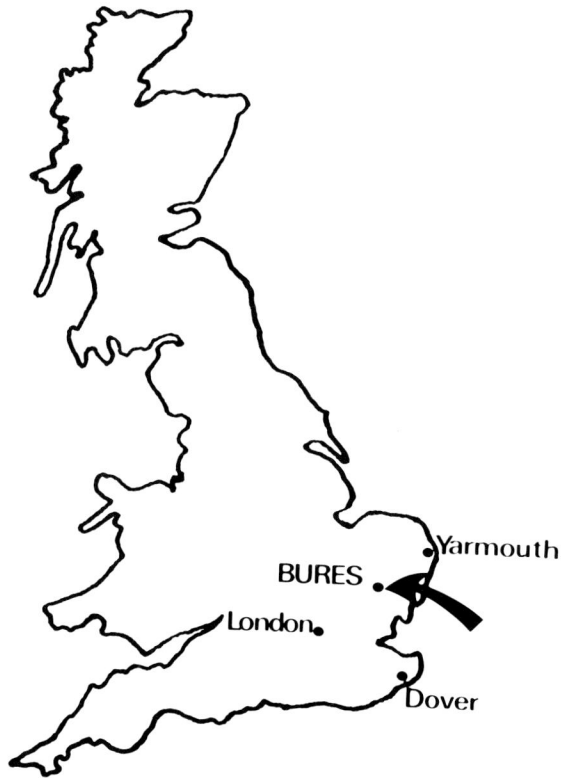

Fig.2.3. The location of the village of Bures in south-east England, from where the earliest transmitters of the gene for Huntington's chorea to the USA were thought to have travelled

persons were possessed by the Devil. This fiend was thought to claw and scratch the suspected individual, leaving conspicuous marks on the skin. This was followed by the application of different ointments which contained, amongst other ingredients, parsley, bat's blood, soot and baby's fat. These persons were thus enabled to fly and participate in the Witches' Sabbath. Submission to the persuasive efforts of the Devil in this way was thought to be tantamount to renouncing God and the Church.

After the death of James I in 1625 his second son, Charles I, succeeded to the throne. The atmosphere of oppression and persecution was intensified and as a result numerous persons, including the three men Jeffers, Nicolas and Wilkie and, by a strange coincidence, the ancestor of George Huntington—a man called Simon Huntington who had lived in Norwich (Fig. 1.4)—emigrated to America. Within a period of 3 years, and from an area spanning only 50 miles, both the earliest transmitters of Huntington's chorea and the progenitor of the man who was to describe definitively the disease left Great Britain for the United States (Critchley 1973).

Unfortunately the views concerning witches were not restricted to England but were also held in the areas where the new immigrants settled, namely Connecticut and Massachusetts. Seven of the female descendants of the three men described by Vessie were regarded as witches. It was believed that the weaving motions, twisting and grimacing of the face and irregular movements of the limbs represented the suffering of Christ during crucifixion and that these persons were so afflicted because they had renounced God (Maltsberger 1961). This was viewed as a most serious offence and

Fig.2.4. A sixteenth-century painting showing the burning of a woman thought to be a witch (Amsterdam 1571)

resulted in 1642 in the first anti-witchcraft laws being passed in New England. Death was suggested as the appropriate punishment (Fig. 2.4).

The fear of witchcraft gained momentum and culminated in the Salem witch-hunt of 1692, which resulted in the death of 19 people (Cotton-Mather 1693; Fig. 2.5). The Massachusetts affair is probably the best documented of witch trials and has intrigued and stimulated numerous authors to ponder its causes and draw present-day equivalents. Arthur Miller's play *The Crucible* is a notable example.

A total of approximately 30 executions took place in New England during the seventeenth century and there is some evidence that persons with Huntington's chorea were among these unfortunates (Maltsberger 1961). Some support for this hypothesis derives from a letter written by the Rev. Samuel Willard, Minister of Groton, Connecticut (Green 1883), concerning one of the grand-daughters of an early settler. He wrote that

this poor and miserable object was observed to carry herself in a strange and unwanted manner... she was violent in bodily motions, leapings, strange agitations... she began by drawing her tongue out of her mouth most frightfully to an extraordinary length and greatnesse and many amusing postures of her bodye.

This description could possibly be a portrayal of the abnormal movements of a person affected with Huntington's chorea.

The Wonders of the Invisible World:

Being an Account of the

T R Y A L S

OF

Several Witches,

Lately Excuted in

N E W - E N G L A N D :

And of several remarkable Curiosities therein Occurring

Together with,

I. Observations upon the Nature, the Number, and the Operations of the Devils.

II. A short Narrative of a late outrage committed by a knot of Witches in *Swede-Land*, very much resembling, and so far explaining, that under which *New-England* has laboured.

III. Some Councels directing a due Improvement of the Terrible things lately done by the unusual and amazing Range of *Evil-Spirits* In *New-England*.

IV. A brief Discourse upon those *Temptations* which are the more ordinary Devices of Satan.

By *COTTON MATHER.*

Published by the Special Command of his EXCELLENCY the Govenour of the Province of the *Massachusetts-Bay* in *New-England*.

Printed first, at *Boston* in *New-England*; and Reprinted at *London*, for *John Dunton*, at the *Raven* in the *Poultry*. 1693.

Fig.2.5. The title page of Cotton Mather's treatise on the witch trials of New England, written in 1693

Vessie has reported that Jeffers, Nicolas and Wilkie all had problems with the law, either on board ship or shortly after their arrival in New England. Wilkie was charged six times for different misdemeanours, including profanity, selling beer without a licence and keeping a disorderly house. Nicolas' dishonesty caused him to be thrown into irons on the ship, while Jeffers was charged with perjury. These sociopathic traits and the mistaken identification of certain of their descendants as witches suggest that these persons may have been suffering from Huntington's chorea. In spite of the fact that Hans and Gilmore (1969) have highlighted certain inaccuracies in Vessie's report, particularly in relation to the alleged presence of chorea in the initial immigrants, it remains accepted that numerous affected individuals are distant descendants of these three men.

Whilst the aforementioned men may be the progenitors of the gene for Huntington's chorea in some affected kindreds, it has been established that other European countries have also contributed to the gene pool for this disorder in America. Hamilton (1908) discovered that some of his patients with chorea in the state of Iowa had not migrated from the eastern seaboard, but had emigrated directly from European countries, including Germany, Ireland and Norway, bringing the gene for Huntington's chorea with them. A similar situation has been reported by Stone and Falstein (1939) for Illinois, where patients were found to be of direct German, English, Scottish, Irish, Scandinavian, Lithuanian or Italian descent.

2.2.2. South Africa

Until recently it was commonly believed that Huntington's chorea was rare in all populations of Africa. However, as a result of nation-wide epidemiological and genealogical studies in South Africa (Hayden et al. 1980a, b) it has been possible to show that the disease is prevalent and was probably introduced to that country over 300 years ago. These findings confirm certain concepts and supplement knowledge concerning the spread of Huntington's chorea around the world.

South Africa offers excellent opportunities for genealogical research because of the systematic recording of the origins and history of its people. Jan van Riebeeck, a Dutchman, arrived at the Cape in 1652 to set up a base for the provision of water and fresh vegetables to the ships of the Dutch East India Company. He kept a meticulous journal, which contained details about many persons and families and paid particular attention to births, deaths, marriages and baptisms. This is an important source of information about events prior to 1665, when church registers were instituted. Most of the immigrants belonged to the Dutch Reformed Church, which also kept records of major events in these people's lives. Both these sources are available for study in their original form.

A most important aid to the investigations was the traditional way in which the early Afrikaner families named their children. The first son was named after his father's father; the second son after his mother's father; the third son after his father. Similarly the first daughter was named after her mother's mother; the second daughter after her father's mother; the third daughter after her mother (Fig. 2.6). Thus, if it was known that the eldest and first-born uncle was named Jacobus Gerhardus Tobias, then the grandfather of the propositus must have been named in the same way. As a result of this tradition missing links in the ancestral lineage of different families have been found.

It was common practice in these kindreds for a bible to be passed down as an heirloom from generation to generation, and a pedigree is sometimes found in the first few pages, with each new addition to the family carefully inscribed, together with his or her progeny. In many instances the bible gave valuable clues to the origins of different families. These

FI

FII

FIII

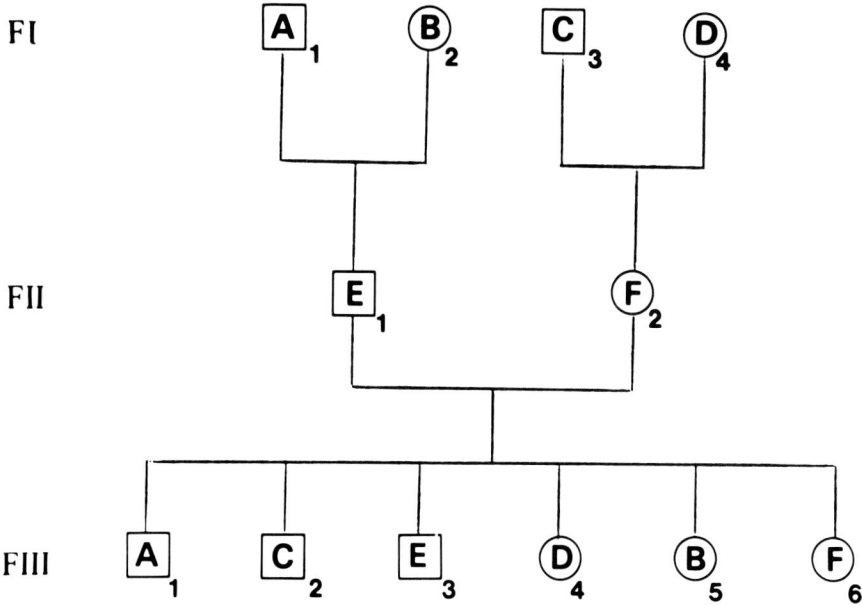

Fig.2.6. The naming of children according to Afrikaner tradition

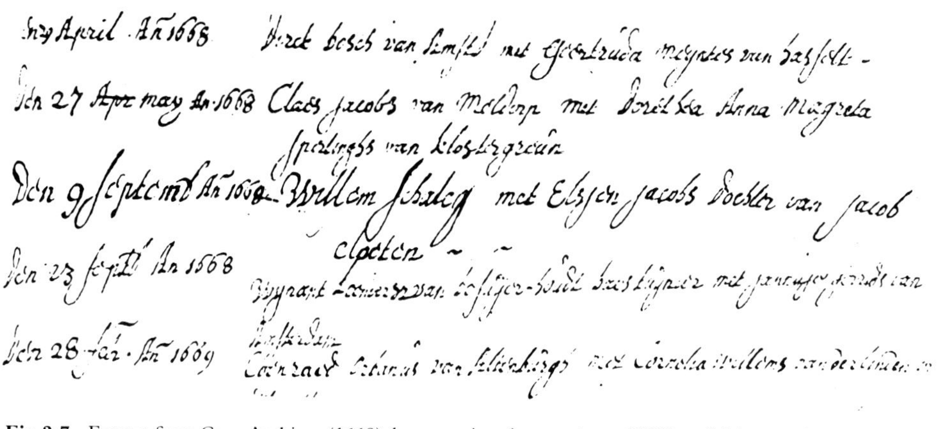

Fig.2.7. Extract from Cape Archives (1668) documenting the marriage of Willem Schlacq with Elsjen Jacobs, the daughter of Jacob Cloeten, on 9 September 1668

sources have greatly facilitated the tracing in South Africa of the ancestral lineage of the gene for Huntington's chorea and other autosomal dominant conditions, such as porphyria (Dean 1971).

Approximately 210 affected individuals in more than 50 families have been found to be ancestrally related through a common progenitor in the seventeenth century. Amongst the 344 persons who arrived aboard the *Dordrecht* at the Cape in 1658 was a man (W.S.vdM.) who was born in Oud-beyerland in the vicinity of Rotterdam. On the 9 September 1668 he married a daughter (E.C.) of one of the earliest settlers in South Africa (Fig. 2.7). The first child of this marriage was a daughter, named Sophia after her maternal grandmother. She married twice and affected individuals have descended from both these marriages (Fig. 2.8). The first definitive evidence of Huntington's chorea has been found in family documents relating to those persons in generation VIII, who have produced numerous affected offspring.

The records of the earliest days at the Cape were searched for any evidence, including a disturbance of handwriting or the presence of antisocial behaviour, which might implicate either Sophia or her parents as heterozygotes for the gene of Huntington's chorea. Even though both her parents signed their wills personally, Sophia (née vdM.) could only sign her own will with a cross (Fig. 2.9). This raised the possibility that she was either affected or illiterate, although the latter seems unlikely as both her parents could write. It is noteworthy that there is on record in Holland a large affected family with a similar name which can be traced to the seventeenth century and which originated from the Rotterdam area (G.W. Bruyn, 1978, personal communication). As W.S.vdM. resided in Oud-beyerland prior to his departure for the Cape, this may be evidence that he was in fact the earliest transmitter of the gene to South Africa.

However, it is clear that there are multiple origins for the gene for Huntington's chorea and of the 74 known unrelated kindreds in South Africa the foreign country of origin has been traced in 34 (Table 2.1).

The abnormal gene has, for example, reached South Africa from France via the island of Mauritius (Hayden 1979). It has been clearly established that all affected persons on this small Indian Ocean island are related and descend from a common ancestor (P. de B.) who was born in the capital of Port Louis in 1864. This man's grandparents, one of whom is thought to have been affected with Huntington's chorea, were of noble descent (Bourbon) and, being dissatisfied with the political situation in France after the French Revolution in 1789, set sail for Mauritius around 1800 aboard the *Mascereynes*. Members of this family migrated to South Africa at the turn of this century to seek better employment opportunities and brought the gene for Huntington's chorea with them (Fig. 2.10). Other relatives of this large kindred at risk for inheriting the disease have emigrated to other islands in the Indian Ocean, including Malagasy (formerly Madagascar), Réunion and the Seychelles, and it is possible that the disorder may also be present in those places.

Table 2.1. Country of origin of earliest identified affected member in 34 South African kindreds

Country of origin	No. of kindreds
England	12
Holland	10
Germany	3
Ireland	2
Scotland	1
Denmark	1
Lithuania (Jewish)	1
Mauritius (French)	1
Cape Verde Islands (Portuguese)	1
Lesotho (African Negro)	1
Lebanon (Semitic)	1
	34

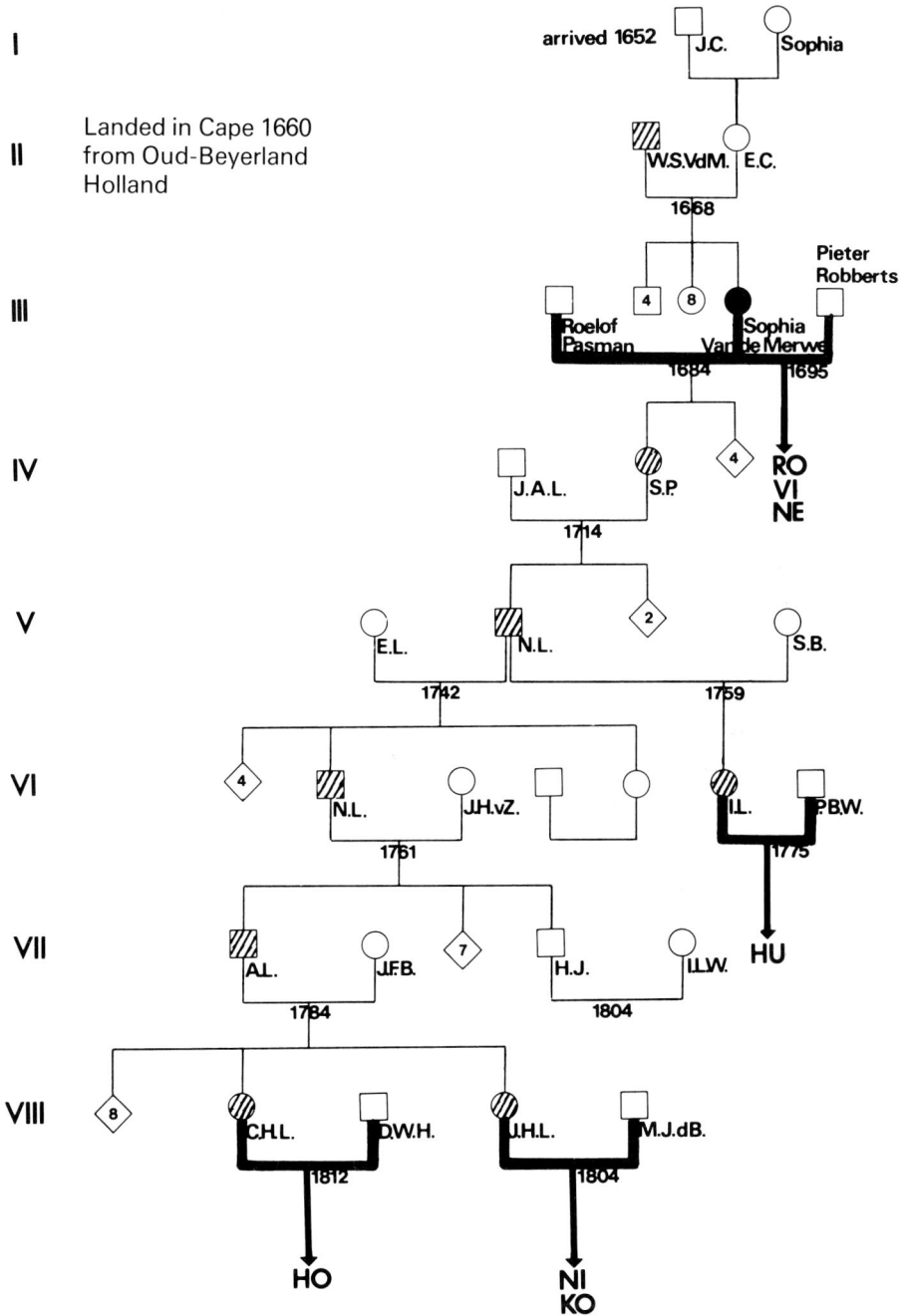

Fig.2.8. Pedigree showing the ancestral relationship of over 200 affected persons to Sophia v.d.M., the most likely earliest transmitter of the gene to South Africa. (From Hayden et al. 1980b.)

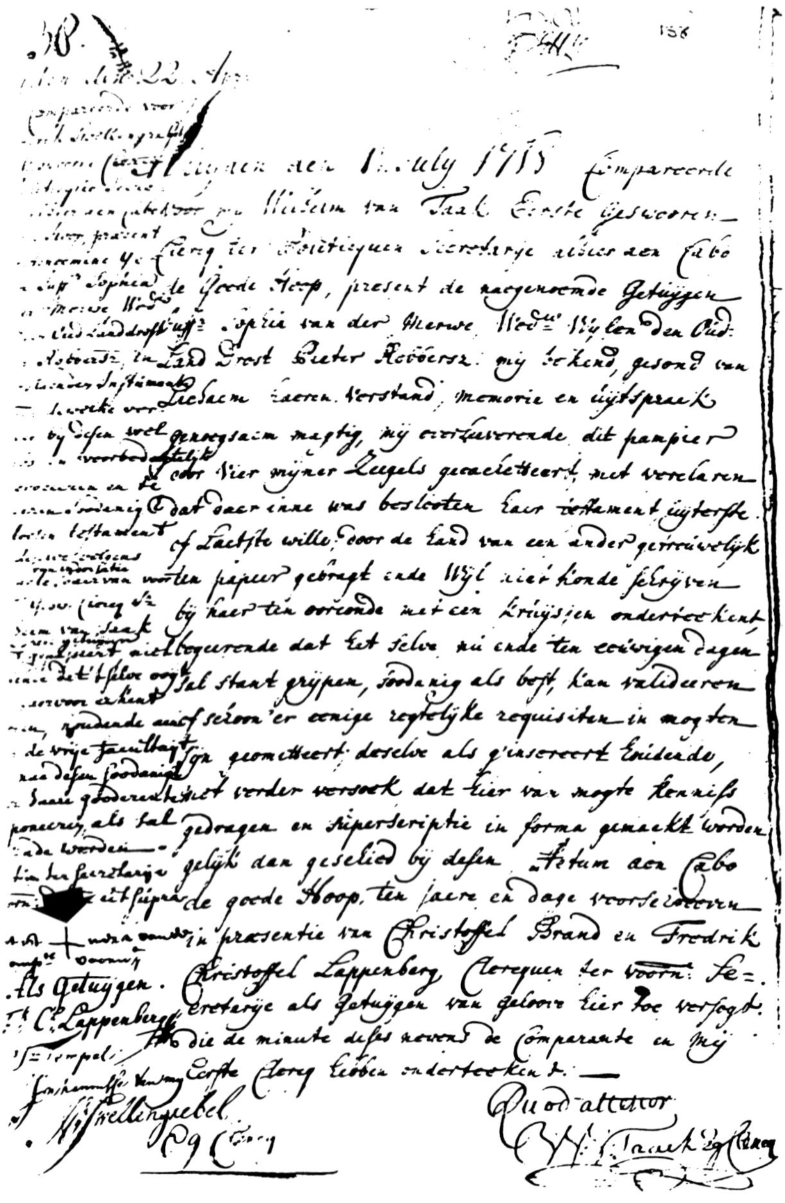

Fig. 2.9. The will of Sophia R. (née vdM) signed with a cross (*arrowed*). (From Cape Archives, CJ 2651)

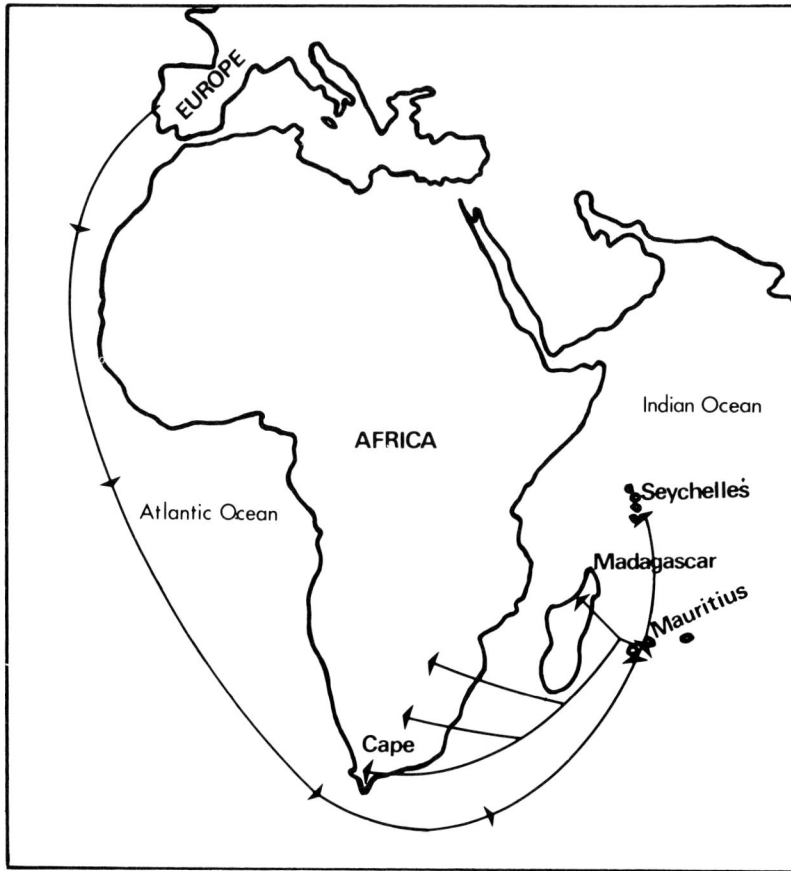

Fig.2.10. The probable migration of the gene for Huntington's chorea to Mauritius in the early nineteenth century and its subsequent spread to neighbouring countries

There have been isolated reports of Huntington's chorea in the African Negroes (Samuels and Gelfand 1978; Glass and Saffer 1979), but they have not included genealogical data. One of the affected African Negroes in South Africa was born in Lesotho, and this suggests that the gene for this disease is also present in that country.

Impressive genealogical studies have been performed in other countries, including Canada (Barbeau and Fullum 1962; Barbeau et al. 1964), Australia (Brothers 1949, 1964), and different parts of Europe including Germany (Entres 1921; Wendt et al. 1960), Belgium (Husquinet 1973) and Norway (Ørbeck and Quelprud 1954). The results of all these investigations suggest that the original source of the gene for Huntington's chorea was north-western Europe.

2.2.3. Canada

The French and the English have been main contributors of the gene to Canada. Barbeau and Fullum (1962) have shown that ancestors of 55 of a total of 75 Canadian kinships under study originally came from the British Isles. Barbeau et al. (1964) have also traced 123 patients with Huntington's chorea of French-Canadian origin who are all ancestrally linked to a single common progenitor who emigrated from France to Montreal in 1645.

2.2.4. Australia

Brothers (1949) has found that the earliest transmitter of the abnormal gene to Australia was a Miss Cundick, who married twice, first to a Mr. Lock and then to a Mr. Viney; affected offspring have been traced to both marriages. She was originally of Huguenot (French) descent. In 1848, after being widowed for the second time, she left her village in the county of Somerset, England, with her 13 children and went to Australia. By 1964 there were 120 affected persons in five generations in Tasmania and 312 patients in Victoria who were ancestrally related to this woman.

2.2.5. The Caribbean

Huntington's chorea is present on the Caribbean island of Trinidad (Beaubrun 1963). Affected individuals are of English, mixed Negro, Spanish and Indian stock. It is of considerable interest that the gene was brought to this particular island by a Brahman priest from the Madras area of India. This implies that the abnormal gene is present in Madras, even though no reports have confirmed this.

2.2.6. The Indian Subcontinent

Most patients with Huntington's chorea in India have been documented in the Punjab, which is over 1500 km from Madras (Chuttani 1955; Singh et al. 1959). However, it seems that the gene is present in other parts of the subcontinent, including Pakistan and Bengal. Affected persons have been traced to their origins in West Pakistan and one patient is on record as having migrated to Bengal (Chuttani 1955). The origin of the gene in these areas remains unknown.

2.2.7. Venezuela

There is a large aggregation of persons with Huntington's chorea in the state of Zulia, along the western shores of Lake Maracaibo in Venezuela. The occurrence of the disease in this locality is closely related to long-standing trade links with Europe. Between 1860 and 1870 dividivi—a local fruit—was exported to Germany, where it was used in the manufacture of high-class dyes. According to legend a Spanish sailor named Antonio Justo Doria abandoned his German merchant ship and entered into free union with a woman from this region. The patients with Huntington's chorea living around Lake Maracaibo are thought to be descendants of this mating (Avila-Giron 1973; see Fig. 7.3).

2.2.8. Moray Firth Area of Scotland

It is intriguing to examine the possible reasons for the enclave of patients with Huntington's chorea in the middle of the Scottish Highlands. This group of persons is unusual in that their names (Patience, MacLeman) are dissimilar from the usual clan names in that area such as MacKenzie, Ross, Munro and others. Sutherland (1969) has suggested two possible explanations for this phenomenon. One is that the gene was

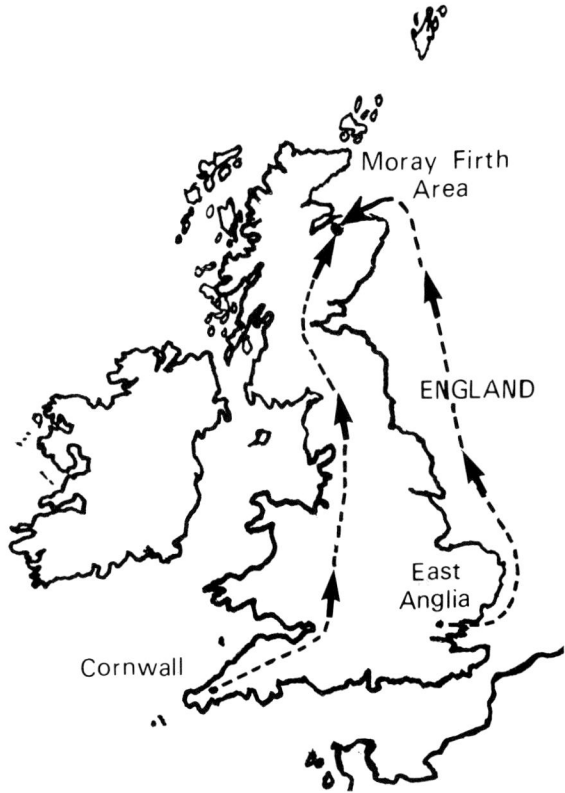

Fig.2.11. Possible origins for the gene of Huntington's chorea in the Moray Firth area of Scotland

brought by a herring fisherman who migrated northwards from East Anglia, England, approximately 300 years ago. An alternative explanation arises from the legend that one of the medieval Earls of Moray married a Cornish heiress whose progeny were supposedly of 'tainted stock'. There is interest in this theory as there is a known aggregation of patients with Huntington's chorea in Cornwall (Bickford and Elisson 1953) (Fig. 2.11).

The disease has until recently been contained in the Moray Firth area by virtue of the fact that these northern fishing families lived in virtually closed communities with little social intercourse outside their own districts. However, current migration patterns will result in the spread of the gene beyond these boundaries.

2.2.9. Japan

Even though the disease is known to have been in Japan for approximately 80 years (Narabayashi 1973) it is not certain whether this is a result of the introduction of the gene at that time or spontaneous mutation. Original migration to Japan occurred from China via Korea. Whilst the disease is known to occur, albeit rarely, in China, it is unlikely that China is the source for the gene in Japan. And though it is known that the Americans established trade links with Japan in the late nineteenth century, there is no reason to postulate that any of those persons were affected with Huntington's chorea. In the absence of a known progenitor the presence of Huntington's chorea in Japan seems likely to be due to spontaneous mutation.

2.3. The Importance of Genealogical Investigations

The investigation of the earliest transmitters of Huntington's chorea to different parts of the world is, in many ways, a study of those countries' histories, their patterns of immigration and their economic relations. Even though there are isolated reports of Huntington's chorea in numerous parts of the world, including Central and South America, Russia and Asia, the origins of the disease in those areas are not clearly established.

  Whilst the investigation of the origins of Huntington's chorea in different parts of the world is of great historical interest, there are also very important medical, diagnostic and socioeconomic implications. The finding that the abnormal gene has been present in a particular country for a prolonged period alone suggests that the disorder may not be uncommon in that area. Furthermore, genealogical studies can alert medical practitioners to maintain a high index of suspicion of Huntington's chorea for persons who present with unexplained psychiatric or neurologic symptomatology in a particular area and who have a surname with which this disorder is known to be associated. Awareness of this situation may reduce the high frequency of initial misdiagnosis in Huntington's chorea.

2.4. The Original Source of the Gene for Huntington's Chorea

The prime fact that emerges from most genealogical investigations is that north-western Europe is the cradle from which the gene for Huntington's chorea has been spread around the world (Fig. 2.12).

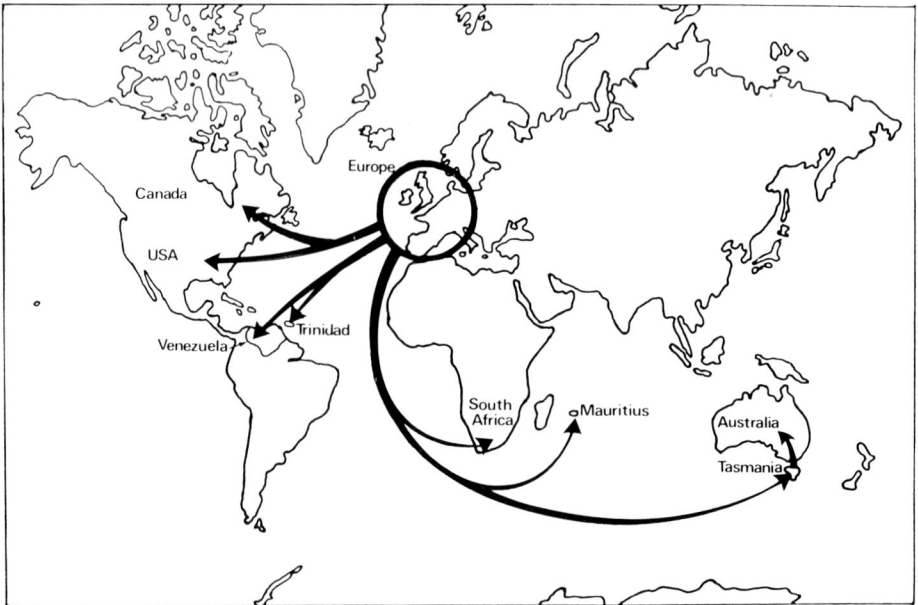

Fig.2.12. The routes of spread of the gene for Huntington's chorea around the world

Persons from affected families were often amongst the earliest pioneering immigrants to different countries, such as the USA and South Africa. Is it possible that one of the reasons for their move was to escape from the dreaded disease and depressing family circumstances, or perhaps the misguided belief that they could avoid this illness by removing themselves physically from their local situation?

Accumulating genealogical evidence implicates France, Germany or Holland as the possible original source of the gene. It is well known that persons of Huguenot descent have taken the disease to America, England and Australia. An interesting but untestable theory is that the gene for Huntington's chorea was present among the Huguenots and that one of the major factors responsible for the spread of the gene around the world was the revocation of the Edict of Nantes on 18 October 1685.

Genealogical studies have clearly established that the disease has been recognised for at least 200 years. It is not known where and in what circumstances in the even more distant past the spontaneous mutation occurred which resulted in the first patient being afflicted with the movement disorder known today as Huntington's chorea. It is feasible that there were persons with Huntington's chorea involved in the dancing mania of the late fourteenth century, but this can never be proved.

References

Avila-Giron R (1973) Medical and social aspects of Huntington's chorea in the state of Zulia, Venezuela. In: Barbeau A, Chase TN, Paulson GW (eds) Huntington's chorea, 1872-1972. Raven Press, New York, pp 261-266

Barbeau A, Fulham G (1962) Origin and migration of Huntington's chorea in Canada. Preliminary report. Can Med Assoc J 87: 1242-1243

Barbeau A, Coiteux C, Trudeau JG, Fullum G (1964) La chorée de Huntington chez les canadiens français: étude préliminaire. Union Med Can 93: 1178-1182

Beaubrun MH (1963) Huntington's chorea in Trinidad. West Indian Med J 12: 39-46

Bickford JAR, Elisson RM (1953) High incidence of Huntington's chorea in the Duchy of Cornwall. J Ment Sci 99: 291-294

Brothers CR (1949) The history and incidence of Huntington's chorea in Tasmania. Proc R Aust Coll Physicians 4: 48-50

Brothers CR (1964) Huntington's chorea in Victoria and Tasmania. J Neurol Sci 1: 405-420

Bruyn GW (1968) Huntington's chorea: historical, clinical and laboratory synopsis. In: Vinken PJ, Bruyn GW (eds) Handbook of clinical neurology. North-Holland Publishing, Amsterdam, pp 298-378

Chuttani PN (1955) Huntington's chorea in the Punjab. J Assoc Physicians India 3: 401

Critchley M (1973) Great Britain and the early history of Huntington's chorea. In: Barbeau A, Chase TN, Paulson GW (eds) Huntington's chorea, 1872-1972. Raven Press, New York, pp 13-17

Davenport CB, Muncey EB (1916) Huntington's chorea in relation to heredity and eugenics. Am J Insan 73: 195-222

Dean G (1971) The porphyrias: a story of inheritance and environment, 2nd ed. Pitman Medical, Tunbridge Wells, Kent, pp 1-176

Dynan NJ (1914) The physical and mental states in chronic chorea. Summary of 19 cases of chronic progressive chorea with the postmortem findings in 8 cases. Am J Insan 70: 589-636

Entres JL (1921) Zur Klinik und Vererbung der Huntington'schen Chorea. Monographien aus dem Gesamtgebiete der Neurologie und Psychiatrie, Heft 27. Springer, Berlin

Glass J, Saffer DS (1979) Huntington's chorea in a Black family. S Afr Med J 56: 685-688

Green SA (1883) Groton in witchcraft times. J Wilson and Son, Cambridge, Mass.

Hamilton AS (1908) A report of 27 cases of chronic progressive chorea. Am J Insan 64: 403-475

Hans MB, Gilmore TH (1969) Huntington's chorea and genealogical credibility. J Nerv Ment Dis 148: 5-13

Hattie WH (1909) Huntington's chorea. Proc Am Med-psychol Assoc (Balt) 16: 171-176

Hayden MR (1979) Huntington's chorea in South Africa. PhD thesis, University of Cape Town

Hayden MR, MacGregor JM, Beighton PH (1980a) The prevalence of Huntington's chorea in South Africa. S Afr Med J 58: 193-196

Hayden MR, Hopkins HC, Macrae M, Beighton P (1980b) The origins of Huntington's chorea in the
    Afrikaner population of South Africa. S Afr Med J 58: 197-200
Husquinet H (1973) Epidemiology and history of Huntington's chorea in Belgium. In: Barbeau A, Chase TN,
    Paulson GW (eds) Huntington's chorea, 1872-1972. Raven Press, New York, pp 245-252
Jelliffe SE (1908) A contribution to the history of Huntington's chorea: a preliminary report. Neurographs 1:
    116-124
Maltsberger JT (1961) Even unto the 12th generation: Huntington's chorea. J Hist Med 16: 1-17
Mather C (1693) The wonders of the invisible world. Boston, New England
Narabayashi H (1973) Huntington's chorea in Japan: review of the literature. In: Barbeau A, Chase TN,
    Paulson GW (eds) Huntington's chorea, 1872-1972. Raven Press, New York, pp 253-260
Ørbeck AL, Quelprud T (1954) Setesdalsrykka (chorea progressiva hereditaria). Nor Videnskap Acad Schr 1:
    1-125
Samuels BL, Gelfand M (1978) Huntington's chorea in a Black Rhodesian family. S Afr Med J 54: 648-651
Singh A, Singh S, Jolly SS (1959) Huntington's chorea: a report of four new pedigrees from Punjab (India).
    Neurol India 7: 7-8
Stevens DL (1976) Huntington's chorea: a demographic, genetic and clinical study. MD thesis, University of
    London, pp 1-338
Stone TT, Falstein EI (1939) Genealogical studies in Huntington's chorea. J Nerv Ment Dis 89: 795-809
Sutherland JM (1969) Geography and diseases of the nervous system. Med J Aust 2: 885-891
Tilney F (1908) A family in which the choreic strain may be traced back to colonial Connecticut. Neurographs
    1: 124-127
Trevelyan GM (1945) History of England. Longmans Green and Co., London (Quoted by Maltsberger 1961)
Vessie PR (1932) On the transmission of Huntington's chorea for 300 years: the Bures family group. J Nerv
    Ment Dis 76: 553-573
Wendt GG, Landzettl I, Solth K (1960) Krankheitsdauer und Lebenserwartung bei der Huntingtonschen
    Chorea. Arch Psychiatr Nervenkr 201: 298-312

3 Epidemiology

Epidemiology is the study of the distribution and causes of health impairment in human populations. In contrast to the clinician, whose primary interest is the individual patient, the epidemiologist focuses on groups of persons. Epidemiological studies are, however, not confined to the determination of disease frequency and may incorporate an evaluation of the measures that need to be taken to improve health.

In spite of the fact that Huntington's chorea has been documented in many parts of the world, formal epidemiological studies have been performed in only a few, and there are still large gaps in our knowledge of the frequency of the disease in Asia, Central and South America, the Middle East and Africa. There has been a decline in the number of epidemiological surveys of Huntington's chorea during the last decade. It would be unfortunate if this trend continued, as such studies offer the only effective method for determining the mutation rate, biological fitness and the effects of public health measures on this disease.

The frequency of Huntington's chorea in any area may be expressed in terms of morbidity and mortality as follows:

1) Morbidity data
   a) *Prevalence*: The total number of affected persons in a defined population (usually per million) at a specific time.
   b) *Incidence*: The total number of newly ascertained affected persons presenting in a given time within a defined population.

2) Mortality data
   a) *Mortality rate*: The number of deaths due to the disease per given time per unit population.

All the figures are expressed as rates in order to facilitate comparison of data.

It would appear from the world-wide distribution of reports of Huntington's chorea that the disease is ubiquitous. The total number of articles concerning this disorder from different countries, including the date of the first publication, is shown in Figs. 3.1 and 3.2. Most articles emanate from the USA, whilst in contrast there have been only 14 from the whole of Africa. The number of publications from a particular country does not necessarily reflect the prevalence of Huntington's chorea but may rather mirror the particular interests of the practising doctor. Nevertheless, it certainly does indicate, at the least, that the disorder is present in that area. These figures have been derived from the *Centennial Bibliography of Huntington's Chorea* (Bruyn et al. 1974), the lists of references in *Index Medicus* up to and including 1979, and the author's private library.

The success of all epidemiological studies depends on many factors, including the completeness of data and the accuracy of medical diagnosis, which in turn depends on the

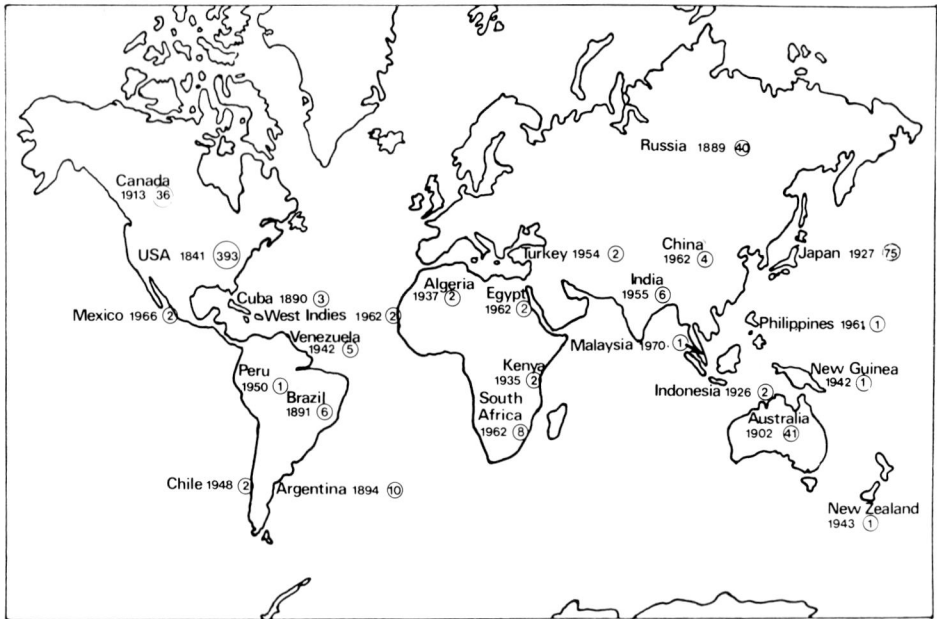

Fig. 3.1. Map of the world showing the total number of articles published on Huntington's chorea in different countries (encircled) and the year of the first publication

experience of the attending clinician. A discussion of the diagnostic criteria for Huntington's chorea thus deserves attention at this stage.

3.1. Diagnostic Criteria

A definitive diagnosis of Huntington's chorea can be made if *each* of the following features is present:

1) A positive family history of typical Huntington's chorea which is consistent with autosomal dominant inheritance.

2) Progressive motor disability, including chorea and/or rigidity of no other obvious cause.

3) Psychiatric disturbance, usually comprising a progressive dementia of no other obvious cause.

Most patients will be diagnosed as a result of having all three features. However, certain anomalous situations may occur. A small proportion of affected persons may have features 1 and 2 without any evidence of psychiatric disturbance. Furthermore, a few persons may present with psychiatric symptomatology without any evidence of motor abnormality. The diagnosis can be accepted if no other cause for these symptoms can be found and, most important, if a positive family history of typical Huntington's chorea can be obtained. When diagnostic difficulty persists, the clinician may choose a deliberate policy of procrastination and use the time to observe the progress of the disorder. This is not a delaying tactic but is indicative of the need for diagnostic precision in Huntington's chorea, in view of the tremendous implications of misdiagnosis.

Fig. 3.2. Map of Europe showing the total number of articles on Huntington's chorea published in each country and the year of the first publication

In some instances it will not be possible to obtain a positive family history for Huntington's chorea. However, a negative genetic history does not necessarily imply that the affected person represents a mutation for the disorder. This may reflect unreliable information about the health of the parents or their early accidental death from unrelated causes before the disease became manifest. The unlikely possibility that the person was not the offspring of the alleged parents, but rather the illegitimate child of an unknown father who carried the gene for Huntington's chorea, should also be considered.

Table 3.1. Reported prevalence of Huntington's chorea ($\times 10^{-6}$ of population) in different countries

Location	Author(s)	Year of publi-cation	Prevalence year	No. of patients	Population	Prevalence (per million)
United Kingdom						
English counties	Critchley	1934				1.9-12.5
London	Minski and Guttmann	1938		43		18
Cornwall	Bickford and Elisson	1953	1950	19	340 941	55.7
Northamptonshire	Pleydell	1954	1954	13	263 000	49
Northamptonshire	Pleydell	1955	1955	17	263 000	65
Northamptonshire	Reid	1960		19	263 000	72
Moray Firth, Scotland	Lyon	1962	1962	5	896	5600
Carlisle, England	Brewis et al.	1966	1961	2		28
NE Metropolitan (Essex region)	Heathfield	1967	1965	81	3 271 000	25
Northamptonshire	Oliver	1970	1968	27	428 000	63
West Scotland	Bolt	1970	1960	154	2 959 600	52
Bedfordshire	Heathfield and Mackenzie	1971		30	427 970	75
East Anglia	Caro	1977	1971	54	584 415	92.4
Leeds/Yorkshire	Stevens	1976	1966	133	3 190 020	41.7
South Wales	Harper et al.	1979	1971	130	1 720 901	75.5
Europe						
Rhineland, Germany	Panse	1942	1933	248	7 690 266	32.2
Switzerland	Zolliker	1949	1900-1930	202		22.5-48.2
Poland	Cendrowski	1964	1964	2		48
West Germany (Kassel region)	Wendt and Drohm	1972	1950	34		27
Haut Vienne, France	Leger et al.	1974	1972	24		70
Belgium (four provinces)	Husquinet	1975	1970	37		16

3.2. Prevalence

The results of the first epidemiological surveys of Huntington's chorea were presented by Critchley (1934) and Sjogren (1936), from the English counties and a parish in northern Sweden respectively. Since that time there have been numerous similar publications, mainly from north-west Europe, Scandinavia, North America, Australia, Japan and South Africa. Whilst there are substantial differences between the reported rates, there is general

Table 3.1. (cont.)   Reported prevalence of Huntington's chorea ($\times 10^{-6}$ of population) in different countries

Location	Author(s)	Year of publication	Prevalence year	No. of patients	Population	Prevalence (per million)
Scandinavia						
North Sweden	Sjogren	1936	1933	18		1440
Iceland	Gudmundsson	1969	1963	5		27
Sweden	Mattsson	1974	1965	362	7 733 853	47
Canada						
Quebec	Barbeau[a]	1966	1966			20-40
Manitoba	Shokeir	1975	1970-1974	162	1 926 942	84
USA						
Minnesota	Pearson et al.	1955	1955	117	3 174 000	54.3
Rochester	Kurland	1958	1955	2		67
Michigan: Whites	Reed et al.	1958	1940	200 ⎫	4 932 652	41.2
Michigan: Blacks	Reed et al.	1958	1940	3 ⎭		15
New York Jews	Myrianthopoulos	1973	1973	±70	2 000 000	35
Long Island and Borough of New York	Korenyi and Whitter	1977	1975	48-250		33.3-203.8
Australia						
Tasmania	Brothers	1949		105	60 344	174
Queensland	Parker	1958	1956	31	1 751 828	23
Victoria	Brothers	1964	1963	138		46
Queensland	Wallace	1972	1969	111	1 751 828	63
Japan						
Aichi district	Kishimoto et al.	1957	1957	13	3 916 922	3.8
South America						
Lake Maracaibo, Venezuela	Avila-Giron	1973	1973	28	±4 000	±7 000
South Africa						
Caucasians	Hayden et al.	1980a	1977	99	4 367 212	22.2
Coloureds (mixed ancestry)				56	2 432 164	21.1
African Blacks				11	16 640 314	0.6

[a] A. Barbeau, 1966, personal communication (quoted by Myrianthopoulos 1973).

agreement that the occidental prevalence of Huntington's chorea is between 30 and 70 affected individuals per million of the population (Table 3.1).

It is noteworthy that there is a general trend towards an increase in the prevalence rates reported in later epidemiological studies compared with those in earlier surveys. This is clearly shown in the surveys undertaken in Queensland in 1958 (Parker) and again in 1972 (Wallace). The prevalence ranges from 23 in the earlier study to 63 in the latter investigation. A similar trend is seen when the results of the first investigations in

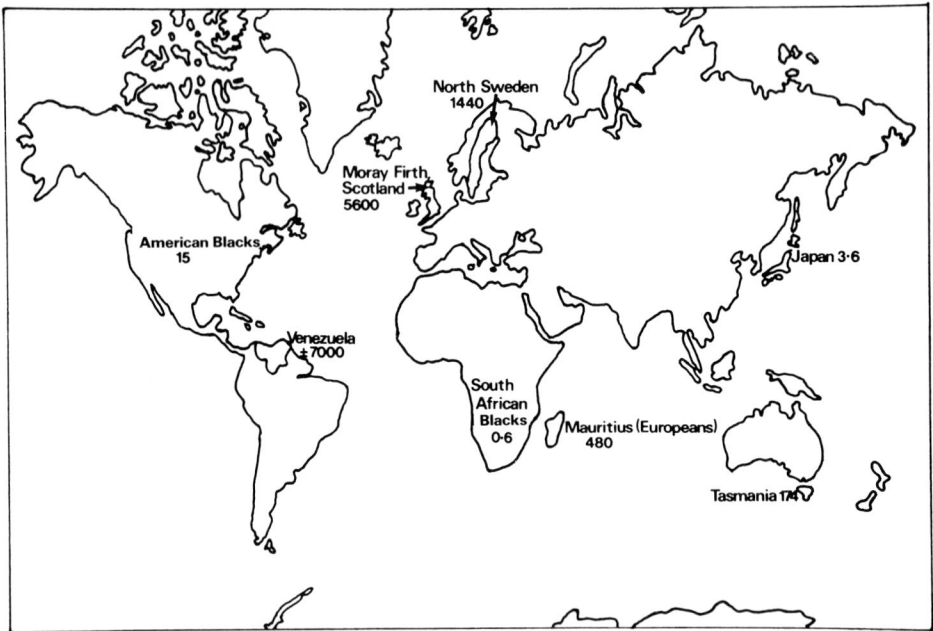

Fig.3.3. Areas of particularly high and low prevalence of Huntington's chorea

Northamptonshire (Pleydell 1954) are compared with the figures presented in 1970. These apparent discrepancies reflect greater accuracy in the collection of data.

Another possibility to be considered, however, is that the true absolute frequency of Huntington's chorea is increasing. This may be particularly true in less developed countries where there is poor dissemination of knowledge about the mode of inheritance of Huntington's chorea and where means of contraception and access to therapeutic termination of pregnancy are less readily available. The possible increase in frequency of the disorder may also be due to its enhanced biological fitness, and this will be discussed further in Chaps. 7 and 8.

3.2.1. Problems in the Comparison of Prevalence Data

Caution must be exercised when comparing prevalence rates from different regions. Prior to ascribing the variation in the morbidity data to racial, ethnic or environmental factors, attention must be focused on the different methods employed in ascertainment of data and the total size of the survey population.

In the earlier surveys (Critchley 1934; Minski and Guttmann 1938) prevalence figures were based solely on the number of affected persons found in mental hospitals. This obviously resulted in underestimation of the true frequency. In later studies attempts have been made to identify all persons suffering from Huntington's chorea in a prescribed geographical area. Reliance has not been placed solely on hospital records and cases have been found by examining information from private physicians, nursing homes, mental institutions and mental health agencies and by extended family tracing. In some of these studies further information has been obtained from professionals involved in the provision of care, such as family practitioners, psychiatrists and neurologists.

Table 3.2. Areas of high prevalence of Huntington's chorea

Location	Author(s)	Rate (per million)	No. of patients	Population size
North Sweden	Sjogren (1936)	1440	18	± 12 500
Tasmania, Australia	Brothers (1949)	174	105	60 344
Moray Firth, Scotland	Lyon (1962)	5600	5	896
Lake Maracaibo, Zulia, Venezuela	Avila-Giron (1973)	± 7000	28	± 4000
Mauritius (Caucasian population)	Hayden (1979)	460	16	± 13000

Because of the lack of standardisation of methods in the earlier surveys, little importance can be attached to the varying prevalence rates they reported. Appropriate and comparable epidemiological criteria have been used in many recent studies and for this reason greater significance can be attached to these reported estimates.

Whilst most prevalence figures from different countries are similar, there are certain areas of particularly high and low frequency of Huntington's chorea (Fig. 3.3).

3.2.2. Areas of High Prevalence: Possible Contributing Factors

The prevalence of Huntington's chorea has been reported to exceed 150 per million of the population in five areas listed in Table 3.2.

A most important factor which must be considered when assessing the prevalence in any region is the total size of the survey population. If this is small, the presence of a few affected individuals artificially raises the prevalence. This is particularly well illustrated in Lyon's study in the Moray Firth area of Scotland, where a total of five affected individuals resulted in the calculated prevalence of 5600 because the total number of inhabitants of this region was under 1000. In all the investigations enumerated above, the small size of the survey population was the major factor in raising the prevalence. For this reason the unusually high figures for the frequency of Huntington's chorea in these relatively isolated communities do not permit inferences to be made concerning the prevalence of the disease in other regions or sections of the community in question.

The genealogy of Huntington's chorea in different parts of the world has been described in Sect. 2.2. Its original sources in Tasmania, Venezuela and Mauritius have been traced to immigrants who arrived in these areas over 100 years ago. Since that time there has been propagation of the gene with a resultant high frequency of Huntington's chorea in these secluded, sparsely populated regions. This illustrates what is known in genetic terms as the 'founder effect'.

In addition to the areas mentioned previously, a high frequency of Huntington's chorea has also been reported in Manitoba, Canada (Shokeir 1975), and East Anglia, England (Caro 1977). In these studies the populations investigated were fairly large, 1 926 942 and 584 415 persons respectively; the prevalence figures were 84 and 92.4. Ascertainment in these surveys was probably near complete due to multiple methods of collection of data. In addition the heightened awareness of Huntington's chorea, particularly in East Anglia, where there has been sustained interest in the disorder for over 50 years, probably led to early diagnosis and improved recognition of patients. The high prevalence rates in these areas are thus not reflections of a small survey population and probably represent fairly accurate estimates of frequency. The reasons for the particular aggregation of Huntington's chorea patients in these regions are at present unknown.

Table 3.3. Areas of low prevalence of Huntington's chorea

Location	Author(s)	Rate (per million)	No. of patients	Population size
Japan	Kishimoto et al. (1957)	3.8	13	3 916 922
American Blacks	Reed et al. (1958)	15	3	207 144[a]
African Blacks (South Africa)	Hayden et al. (1980a)	0.6	11	16 640 314

[a] Calculated from data provided.

Preliminary results of a pilot survey in the Liguria province of northern Italy (C. Loeb, 1980, personal communication) show that this may be yet another area with an unusually raised prevalence rate for Huntington's chorea. However, this finding awaits confirmation in a large-scale epidemiological survey.

3.2.3. *Areas of Low Prevalence: Possible Contributing Factors*

Huntington's chorea appears to be particularly uncommon in Japan, and among the African and American Blacks (Table 3.3).

Huntington's chorea in Japan is one-tenth as common as in most Western countries. Reed et al. (1958) found that this disorder is also 3 times less frequent in American Blacks than in American Whites. This latter study was performed in Michigan, where only three affected Blacks were accurately documented. More recent data from the large medical care system of the Veterans Administration Hospitals in the USA have confirmed that the rate of Huntington's chorea in US Blacks is less than half that of US Whites (Kurtzke et al. 1977). The disease is also approximately 40 times less frequent in African Blacks than in South African Whites. It is noteworthy that other disorders of the basal ganglia, including Parkinsonism, are also rare in the African Black population (Wasserman 1974).

All the figures noted above were obtained during extensive investigations and it is certain that the low frequency of Huntington's chorea in these peoples is not the result of underestimation but represents a real racial difference in prevalence. The explanation for this geographic variation in frequency is closely related to an understanding of the ethnic origins of these peoples on the one hand and the history of Huntington's chorea on the other.

It is likely that the higher frequency of this disorder in the American as opposed to the African Black reflects a much larger European genetic component in the former compared with his African counterpart, who has few Caucasian genes.

All genealogical studies indicate that emigration from north-western Europe was primarily responsible for the spread of the gene for Huntington's chorea around the world. A diminished frequency of the disorder would then be expected in those races who have their origins outside Europe. The low prevalence in the North American and South African Blacks and the Japanese is in keeping with this hypothesis.

3.3. The Epidemiology of Juvenile Huntington's Chorea

Juvenile Huntington's chorea refers to persons who present with definite signs and symptoms before the age of 20. The proportion of juvenile patients generally varies between 1% and 10% in different series (Table 3.4). The highest frequency of this precocious form of the disorder has been reported in the population of mixed ancestry in

Table 3.4. Number of juvenile patients expressed as a percentage of the total survey population

Author(s)	Childhood (0-9) No.	Childhood (0-9) %	Adolescent (10-19) No.	Adolescent (10-19) %	Total No.	Total %	Location
Davenport and Muncey (1916)	3	2.2	2	1.4	5	3.6	USA
Spillane and Phillips (1937)	–	–	1	4.7	1	4.7	Wales
Panse (1942)	9	2.0	22	4.9	31	6.9	Germany
Reed et al. (1958)	–	–	12	5.8	12	5.9	USA
Brothers (1964)	4	1.9	11	5.3	15	7.2	Australia
Cameron and Venters (1967)	2	0.9	9	3.9	11	4.8	Scotland
Heathfield (1967)	–	–	2	2.9	2	2.9	England
Oliver (1970)	4	3.5	7	6.1	11	9.6	England
Mattsson (1974)					13	8.0	Sweden
Stevens (1976)	1	0.3	2	0.7	3	1.0	England
Hayden et al. (1981)							South Africa
	6	2.7	11	5.0	17	7.7	Total
	3	2.0	3	2.0	6	4.0	Whites
	3	4.3	8	11.4	11	15.7	Mixed ancestry

South Africa (Hayden et al. 1981). In contrast to the European group of that country, in which the population of juvenile patients is 4%, in the community of mixed ancestry 15.7% of persons with Huntington's chorea have onset before the age of 20.

There are no obvious environmental influences which account for this finding and it is most likely that genetic factors are important in causing the ethnic differences in frequency of juvenile Huntington's chorea in that region. The Whites are predominantly of north-western European origin and it is therefore not surprising that their proportion of juvenile patients is similar to that found in European countries. The population of mixed ancestry, however, has a very different genetic constitution, originally resulting from intermingling between Whites, Khoisan (Bushmen and Hottentots) and Malaysian slaves in the seventeenth and eighteenth centuries. This was brought about because the European immigration to the Cape during the earliest years of the settlement was predominantly male, and many of these men who failed to find wives in their own ethnic group entered into unions with Khoisan or Malaysian women.

It has been mentioned in Sect.2.2.2 that the gene for Huntington's chorea was probably introduced to South Africa and to the population of mixed ancestry by a Dutch settler who arrived in 1660. It will be shown later, in Sect. 7.5.1, that the father is the transmitter of the gene to affected juveniles 3 to 4 times more frequently than the mother. It could be suggested that there might be some link between the high frequency of juvenile patients in the population of mixed ancestry and the predominant White male contribution to their genetic origins. This hypothesis could be tested by investigating other populations of mixed ancestry in different parts of the world where it is known that an affected male first transmitted the disease to the group and where males of north-west European descent made a major contribution to their origins.

Table 3.5.   Annual incidence per million population: a review

Author(s)	Location	Incidence per million	Expected no. of new patients/year
Reed et al. (1958)	Michigan (USA): Whites	2.59 ⎫	12.6
	Michigan (USA): Negroes	0.94 ⎭	
Brothers (1964)	Victoria, Australia	3.75	11.75
Bolt (1970)	Scotland (West)	3.63	10.7
Stevens (1976)	England (Leeds area)	3.06	9.76
Hayden (1979)	South Africa: Whites	1.5	6.5
	South Africa: mixed ancestry	1.7	4.1
	South Africa: Blacks	0.04	0.6

3.4. Incidence

The incidence of Huntington's chorea refers to the total number of newly affected persons who present within a given time period. In a stable situation, assuming the prevalence and duration of this chronic disease to be constant,

$$\text{Incidence} = \frac{\text{Prevalence}}{\text{Duration}}$$

The prevalence is expressed per million of the population, the duration defined in terms of years, and therefore the incidence is denoted per million of the population per year. The total number of persons who may present per year is equal to the incidence multiplied by the population (expressed in millions).

3.4.1. International Comparison

The annual incidence of Huntington's chorea in different parts of the world is shown in Table 3.5. The mean incidence in populations of European ancestry is higher than that seen in the African or American Blacks. These findings are consistent with the prevalence data presented previously.

It is always difficult to assess the reliability of epidemiological data. One measure of accuracy is to compare the actual number of patients presenting in 1 year with the expected number of patients in that same area in the given time. In other words, one is comparing the actual incidence with the expected incidence, which can be determined from prevalence and duration data. In most instances the actual figures will be higher than the expected results.

3.5. Mortality Data

The importance of mortality data is that they may reflect morbidity patterns and are another index of the minimum frequency of disease in a specific community. In contrast to prevalence rates, which are derived from large population surveys, mortality data are calculated from government notification of death rates, which is based on official death certificates completed by the attending medical practitioners.

Table 3.6. International comparison of numbers of deaths coded 331.0 as underlying cause and death rates per million population per year (1968-1975)

Location	Males	Females	Total	Rate
USA (Whites)	704	855	1559	1.25
USA (Blacks)	34	34	68	0.37
USA (whole population)	738	889	1627	1.14
Sweden	41	42	83	1.71
Denmark	31	33	64	1.84
Japan	42	57	99	0.13
England and Wales[a]	489	519	1008	1.5
South Africa (Whites)[a]	18	19	37	1.03
South Africa (mixed ancestry)[a]	3	6	9	0.46
South Africa (African Blacks)[a]	1	2	3	0.02

Adapted from Kurtzke (1979).

[a]Code 331, total.

The underlying cause of death is coded according to a four-digit number, which represents a specific diagnosis within the International Statistical Classification of Diseases (ISC). Huntington's chorea was assigned the number 331.0 in the eighth revision of the ISC, which was in effect from 1968 until the end of 1978. Other components of code 331 include diseases of rare occurrence and thus the total sum of deaths coded 331 is very close (albeit somewhat higher) to the true number of persons dying from Huntington's chorea. This is most opportune as three-digit codes are used for the classification of death data from most countries, excluding Scandinavia where more specific four-digit coding is the accepted convention.

The ninth revision of the ISC came into effect on 1 January 1979 and has coded Huntington's chorea as 333.4. Unfortunately the rubric of 333 now includes many other diseases which occur with greater frequency than those which were previously listed in association with Huntington's chorea. Thus this three-digit number will be of less value when assessing the number of deaths from Huntington's chorea in the future.

3.5.1. *International Comparison*

The death rates due to Huntington's chorea for different countries are shown in Table 3.6. The figures for north-west Europe and for the US Whites are similar, ranging between 1 and 2 deaths per million of the population per year. The higher death rates in Scandinavia do not reflect a truly increased morbidity or mortality for Huntington's chorea in that area but rather more accurate and meticulous collation of data in smaller populations. The low death rates for the Japanese, US Blacks and African Blacks parallel and confirm the findings of low prevalence data in these population groups.

Death certificates in most countries record both the underlying cause of death and the contributory cause of death. However, mortality data are derived from only the former notification. Thus if the affected person died from pneumonia and Huntington's chorea was only noted as the associated cause, this patient would not be included under the digit for Huntington's chorea. If the secondary (contributory) death rates are included, this will result in an increase over the primary (underlying) death data by almost half (Kurtzke 1979).

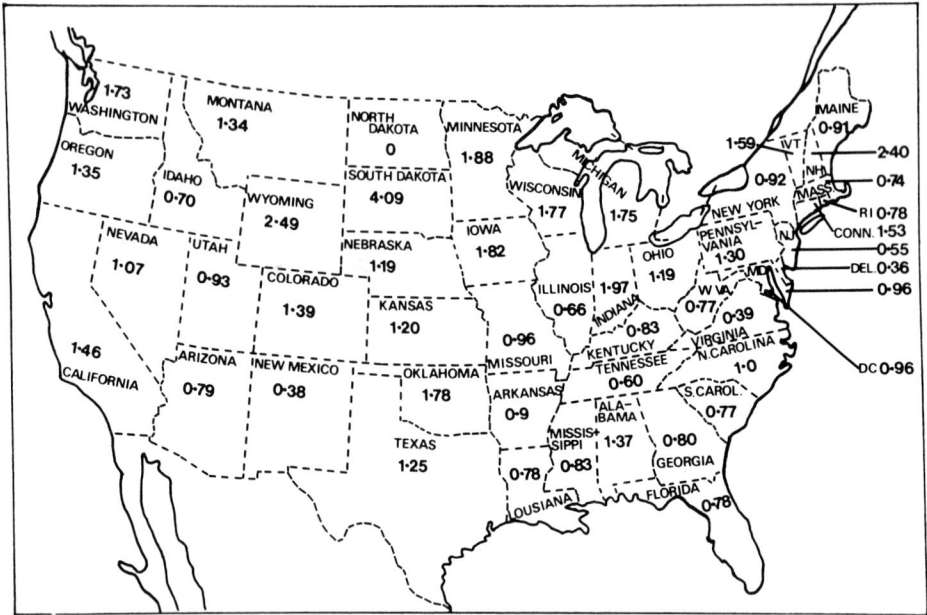

Fig.3.4. Regional comparison of mortality rates of Huntington's chorea (expressed per million) in the USA between 1968 and 1974. (Adapted from Hog et al. 1979.) Alaska and Hawaii, which are not depicted, had no deaths coded to this cause during this period

Mortality data for Huntington's chorea between 1968 and 1974 in the USA have been carefully analysed by Hogg et al. (1979). A regional comparison of death rates in different states is shown in Fig. 3.4. Major variations in mortality data are apparent, the highest rates being reported from South Dakota, Wyoming and New Hampshire, whilst North Dakota, Alaska and Hawaii had no deaths coded for Huntington's chorea. It is feasible to assume that all the different states had the same problems with regard to the calculation of death rates, such as incomplete reporting, errors in coding and possible misdiagnosis. Thus the differing mortality data may reflect true variation in morbidity patterns and may highlight those areas in special need of resources for the management of patients and their families.

Analysis of the Registrar General's Statistics for the United Kingdom shows a marked increase in the overall number of deaths attributed to Huntington's chorea in the last two decades (Scoones 1980). This is closely related to greater awareness of the disorder, improved diagnosis and more efficient recording, but could also be a further indication of an increase in the frequency of Huntington's chorea in Great Britain (Fig. 3.5).

The most accurate assessment of the frequency of Huntington's chorea will be obtained in a survey of a large population in which multiple methods of ascertainment are used. It would appear that mortality data are particularly unreliable parameters for estimating the importance of this disease. Until such time as the social stigma of Huntington's chorea has been overcome, all statistics concerning its epidemiology will remain deceptively low. Stated differently, the completeness of collection of data and thus the estimated frequency of Huntington's chorea in any area is inversely proportional to the social stigma attached to it.

DEATHS

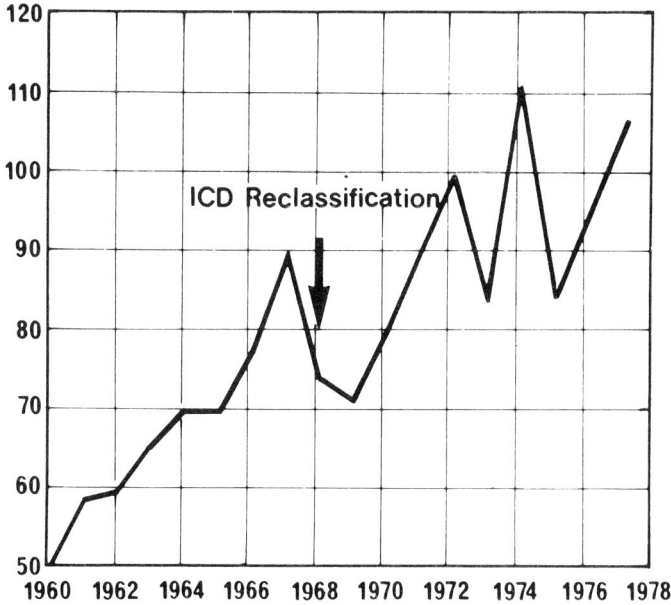

Fig.3.5. The increase in the reported number of deaths due to Huntington's chorea in the United Kingdom in recent years. (From Scoones 1980)

References

Avila-Giron R (1973) Medical and social aspects of Huntington's chorea in the state of Zulia, Venezuela. In: Barbeau A, Chase TN, Paulson GW (eds) Huntington's chorea, 1872-1972. Raven Press, New York, pp 259-266

Bickford JAR, Elisson RM (1953) High incidence of Huntington's chorea in the Duchy of Cornwall. J Ment Sci 99: 291-294

Bolt JM (1970) Huntington's chorea in the west of Scotland. Br J Psychiatry 116: 259-270

Brewis M, Poskanzer DC, Rolland C, Miller W (1966) Neurological disease in an English city. Acta Neurol Scand [Suppl 24] 42: 1-89

Brothers CRD (1949) The history and incidence of Huntington's chorea in Tasmania. Proc R Aust Coll Physicians 4: 48-50

Brothers CR (1964) Huntington's chorea in Victoria and Tasmania. J Neurol Sci 1: 405-420

Bruyn GW, Baro F, Myrianthopoulos NC (1974) Centennial bibliography of Huntington's chorea, 1872-1972. Leuven University Press and Martinus Nijhoff, Louvain and The Hague: pp 1-314

Cameron D, Venters GA (1967) Some problems in Huntington's chorea. Scott Med J 12: 152-156

Caro A (1977) The prevalence of Huntington's chorea in an area of East Anglia. J R Coll Gen Pract 27: 41-45

Cendrowski W (1964) Niektore dane o geografii plasawicy dxiedzicznej. (Some remarks on the geography of hereditary chorea.) Neurol Neurochir Psychiatr Pol 14: 63-66

Critchley M (1934) Huntington's chorea and East Anglia. J State Med 42: 575-587

Davenport CB, Muncey EB (1916) Huntington's chorea in relation to heredity and eugenics. Am J Insan 73: 195-222

Gudmundsson KR (1969) Prevalence and occurrence of some rare neurological diseases in Iceland. Acta Neurol Scand 45: 114-148

Harper PS, Tyler A, Walker DA, Newcombe RG, Davies K (1979) Huntington's chorea: the basis for long-term prevention. Lancet II: 346-349

Hayden MR (1979) Huntington's chorea in South Africa. PhD thesis, University of Cape Town

Hayden MR, MacGregor JM, Beighton PH (1980) The prevalence of Huntington's chorea in South Africa. S Afr Med J 58: 193-6

Hayden MR, MacGregor JM, Saffer DS, Beighton PH (1981) On the high frequency of juvenile Huntington's chorea in South Africa. J Med Genet (in press)

Heathfield KWG (1967) Huntington's chorea: investigation into the prevalence of this disease in the area covered by the North East Metropolitan Regional Hospital Board. Brain 90: 203-232

Heathfield KWG, MacKenzie ICK (1971) Huntington's chorea in Bedfordshire. Guy's Hosp Rep 120: 295-309

Hogg JE, Massey EW, Schoenberg BS (1979) Mortality from Huntington's disease in the United States. In: Chase, TN, Wexler NS, Barbeau A (eds) Advances in neurology, vol 23. Raven Press, New York, pp 27-35

Husquinet H (1975) La chorée de Huntington, maladie héréditaire grave: conditions de sa disparition. Rev Med Liège 30 7: 228-232

Kishimoto K, Nakamura M, Sotokawa Y (1957) Population genetics study of Huntington's chorea in Japan. Annu Rep Res Inst Environ Med 9: 195-211

Korenyi C, Whitter JR (1977) Huntington's disease in New York State. New York J Med 77 1: 44-45

Kurland LT (1958) Descriptive epidemiology of selected neurologic and myopathic disorders with particular reference to a survey in Rochester, Minnesota. J Chron Dis 8: 378-418

Kurtzke JF (1979) Huntington's disease: mortality and morbidity data from outside the United States. In: Chase TN, Wexler NS, Barbeau A (eds) Advances in neurology, vol 23. Raven Press, New York, pp 13-25

Kurtzke JF, Anderson VE, Beebe GW, Elston RC, Higgins I, Hogg J, Kurland L, Muenter M, Myrianthopoulos N, Reed TE, Schoenberg B, Schull WJ, Li CC (1977) Report of workgroup on epidemiology, biostatistics and population genetics. In: Commission for the Control of Huntington's Disease and its Consequences vol III/1. DHEW publication (NIH) 718-1503. US Government Printing Office, Washington, DC, pp 133-236

Leger JM, Ranouil R, Vallat JN (1974) Huntington's chorea in Limousin: statistical and clinical study. Rev Med Limoges 5: 147-153

Lyon RL (1962) Huntington's chorea in the Moray Firth area. Br Med J i: 1301-1306

Mattsson B (1974) Clinical, genetic and pharmacological studies in Huntington's chorea. UMEA University Medical Dissertations 7. UMEA, Sweden, pp 21-51

Minski L, Guttmann E (1938) Huntington's chorea: study of 34 families. J Ment Sci 84: 21-96

Myrianthopoulos NC (1973) Huntington's chorea: the genetic problem five years later. In: Barbeau A, Chase TN, Paulson GW (eds) Huntington's chorea, 1872-1972. Raven Press, New York, pp 150-152

Oliver JE (1970) Huntington's chorea in Northamptonshire. Br J Psychiatry 166: 241-253

Panse F (1942) Die Erbchorea: eine klinische-genetische Studie. Samml Psychiatr Neurol Einzeldarst 18. Thieme, Leipzig

Parker N (1958) Observations on Huntington's chorea based on a Queensland survey. Med J Aust I: 251-259

Pearson JS, Petersen MC, Lazarte JA, Blodgett HE, Kley IB (1955) An educational approach to the social problem of Huntington's chorea. Proc Mayo Clin 30: 349-357

Pleydell MJ (1954) Huntington's chorea in Northamptonshire. Br Med J ii: 1121-1128

Pleydell MJ (1955) Huntington's chorea in Northamptonshire. Br Med J ii: 889

Reed TE, Chandler JH, Hughes EM, Davidson RT (1958) Huntington's chorea in Michigan: demography and genetics. Am J Hum Genet 10: 201-225

Reid JJ (1960) Huntington's chorea in Northamptonshire. Br Med J ii: 650

Scoones T (1980) Huntington's chorea. Office of Health Economics Publication 67. OHE, London, pp 1-35

Shokeir MH (1975) Investigations on Huntington's disease in the Canadian prairies. II. Fecundity and fitness. Clin Genet 7: 349-353

Sjogren T (1936) Vererbungsmedizinische Untersuchungen über Huntington's Chorea in einer schwedischen Bauernpopulation. Z Menschl Vererb-Konstit-Lehre 19: 131-165

Spillane J, Phillips R (1937) Huntington's chorea in South Wales. Q J Med 6: 403-423

Stevens DL (1976) Huntington's chorea: a demographic, genetic and clinical study. MD thesis, University of London, pp 1-338

Wallace DC (1972) Huntington's chorea in Queensland: a not uncommon disease. Med J Aust II: 1275-1277

Wasserman HP (1974) Ethnic pigmentation: historical, physiological and clinical aspects. Excerpta Medica, Amsterdam, pp 254-255

Wendt GG, Drohm D (1972) Krankheitsdauer und Lebenserwartung bei der Huntingtonschen Chorea. Arch Psychiatr Nervenkr 201: 298-312

Zolliker A (1949) Die Chorea Huntington in der Schweiz. Eine Thurgauer Sippe mit Chorea Huntington. Schweiz Arch Neurol Psychiatr 64: 448-457

4 Natural History

A review of a disorder is incomplete without a discussion of its natural history, which is essentially a study of the behaviour of the disease process in a population. The two most important measurable points in the natural history are the age at onset and age at death. The duration of the disease is equal to the latter figure minus the former.

4.1. Age at Onset

Considerable confusion has arisen over what constitutes onset of Huntington's chorea. Persons who are heterozygous for this disease clearly have the gene from conception. Some might argue that the disease therefore has its onset at conception, although it only becomes clinically manifest many years later. In some epidemiological surveys this issue has been avoided by including results of mean ages at onset without accurately defining the criteria used. Without these specifications, however, comparison of results of different studies has doubtful significance.

A broad definition of age at onset is 'the first time any signs or symptoms appear which are either neurological or psychiatric and represent a permanent change from the normal state'. However, even with this interpretation, accurate estimation of the age at onset may still be most difficult. Doctors are seldom witness to the earliest signs and symptoms of Huntington's chorea and assessment of these insidious changes often depends on the patients themselves or their families. Denial of Huntington's chorea in its early phases is commonplace, with acknowledgement of the presence of the disorder only in its more advanced form and resultant overestimation of age at onset. In contrast, other families are over-zealous to ascribe any restlessness or clumsiness to the disease and in this way normal behaviour may be misconstrued and thereby age at onset miscalculated.

4.1.1. International Comparison

The mean ages at onset of Huntington's chorea in reports from different parts of the world are shown, chronologically by year of report, in Table 4.1. A striking feature is the marked variation in this parameter, which ranges from 33.8 (Brackenridge 1971) to 51.59 years (Lyon 1962). Adequate definitions of the criteria used to denote onset of the disease are notable omissions from many studies (Bickford and Elisson 1953; Cameron and Venters 1967; Davenport and Muncey 1916; Heathfield 1967; Lyon 1962; Oliver 1970; Shokeir 1975). Reed et al. (1958) considered the beginning of the disorder to be the time of appearance of chorea and thereby disregarded pre-existing psychiatric symptoms. Other authors (Bolt 1970; Brothers 1964; Hayden 1979; Mattsson 1974; Panse 1942;

Table 4.1.   Age at onset (AaO) of Huntington's chorea: an international review

Location	Author(s)	Year	No. of patients	Mean AaO
USA	Davenport and Muncey	1916	138	37.8
Germany	Entres	1921	323	38.1
Literature review	Bell	1934	460	35.5
Sweden	Sjogren	1936	48	41.4
South Wales	Spillane and Phillips	1937	21	41.2
London	Minski and Guttmann	1938	43	42.4
Rhineland	Panse	1942	446	37.2
Switzerland	Zolliker	1949	120	40.5
Cornwall, England	Bickford and Ellison	1953	21	42.8
Norway	Ørbeck and Quelprud	1954	109	39.8
Michigan, USA	Reed et al.	1958	204	35.3
Moray Firth, Scotland	Lyon	1962	41	51.59
Victoria, Australia	Brothers	1964	206	37.2
England	Heathfield	1967	69	44.2
Scotland	Cameron and Venters	1967	230	43.2
France	Petit	1969	125	39.3
England	Oliver	1970	115	36.4
Scotland	Bolt	1970	265	42.7
Literature review	Brackenridge	1971	344	33.8
Belgium/France/Holland	Husquinet et al.	1973	751	41
Sweden	Mattsson	1974	162	37.1
Canada	Shokeir	1975	162	40.54
Leeds, England	Stevens	1976	298	42.95
South Africa	Hayden	1979	219	34.93

Stevens 1976) have defined age at onset as the first time at which any symptoms of the disorder appear. From the foregoing it is evident that the different criteria used in the varying definitions of this parameter account for some of the marked differences in the reported age at onset.

The number of persons presenting with Huntington's chorea in each decade of life, as reported in different investigations, is shown in Table 4.2. The total in each age cohort is also presented as a percentage of the sum of the whole group. It is evident that Huntington's chorea may appear at any stage of life. Whilst the majority of affected persons have onset in the fourth decade (28.3%), only 9 persons of a total of 2699 (Table 4.2) showed first symptoms after the age of 70, and 4.7% had onset before the age of 20. The range of age at onset in 802 patients of Wendt and Drohm's study in Germany in shown in Fig. 4.1. In the South African investigation (Fig. 4.2) the whole curve is shifted to the left, reflecting the general trend to lower age at onset.

In this context it is obvious that a high proportion of persons with juvenile Huntington's chorea lowers the estimated mean age at onset of a survey population. It is of interest that where the number of juvenile patients is low (Heathfield 1967: 2.9%; Stevens 1976: 1%) the mean age at onset is high, namely 44.2 and 42.95 years respectively. Conversely, a higher proportion of juvenile cases (Reed et al. 1958: 5.9%; Brothers 1964: 7.2%; Oliver 1970: 9.6%) seems to parallel a decrease in the mean age at onset in these studies (35.3, 37.2 and 36.4 years respectively). However, this hypothesis does not resolve all the variations between different investigations. Davenport and

Table 4.2. Age at onset in decades: a review

Age	Davenport and Muncey (1916) No.	%	Spillane and Phillips (1937) No.	%	Panse (1942) No.	%	Reed et al. (1958) No.	%	Wendt et al. (1959) No.	%	Brothers (1964) No.	%	Cameron and Venters (1967) No.	%	Heathfield (1967) No.	%	Oliver (1970) No.	%	Stevens (1976) No.	%	Hayden (1979) No.	%	Total No.	%
0- 9	3	2.2	–	–	9	2.0	–	–	1	0.2	4	0.9	2	0.9	–	–	4	3.5	1	0.3	6	2.7	30	1.1
10-19	2	1.4	1	4.7	22	4.9	12	5.9	16	2.2	11	5.3	9	3.9	2	2.9	7	6.1	2	0.7	11	5.0	95	3.6
20-29	16	11.6	2	9.5	105	23.5	42	20.6	82	10.8	39	18.9	21	9.1	4	5.8	17	14.8	18	6.2	56	25.6	402	14.9
30-39	50	36.2	4	19.1	129	28.9	77	37.7	170	22.3	65	31.6	60	26.1	18	26.1	31	27.0	82	28.4	77	35.2	763	28.3
40-49	35	25.4	8	38.2	113	25.4	61	29.9	276	36.2	49	23.8	70	30.4	24	34.8	34	29.5	107	37.0	43	19.6	820	30.4
50-59	25	18.1	5	23.8	53	11.9	10	4.9	180	23.6	30	14.6	43	18.7	19	27.5	18	15.7	64	22.1	24	11.0	471	17.4
60-69	7	5.1	1	4.7	13	2.9	2	1.0	35	4.5	8	3.9	23	10.0	2	2.9	4	3.4	12	4.3	2	0.9	109	4.0
70	–	–	–	–	2	0.5	–	–	2	0.2	–	–	2	0.9	–	–	–	–	3	1.0	–	–	9	0.3
Total	138	100	21	100	446	100	204	100	762	100	206	100	230	100	69	100	115	100	289	100	219	100	2699	100

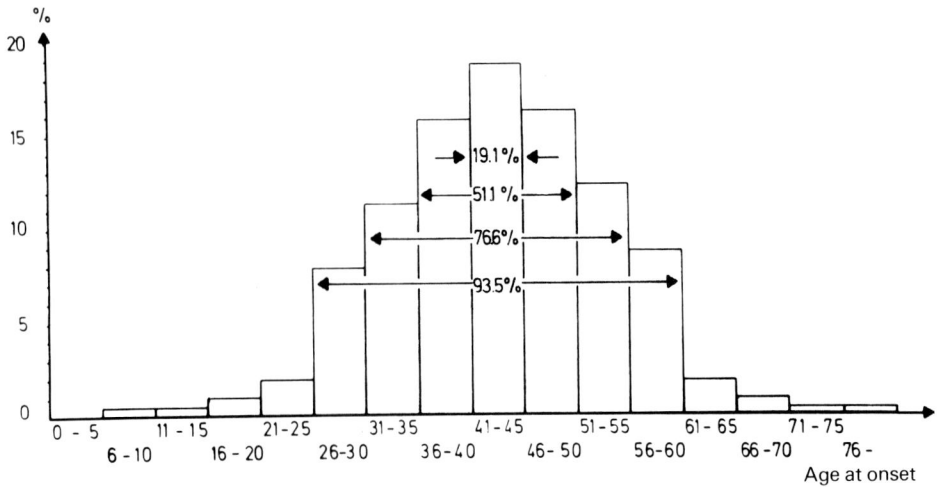

Fig.4.1. Distribution of ages at onset in 802 patients in Wendt and Drohm's survey. (From Vogel and Motulsky 1979.)

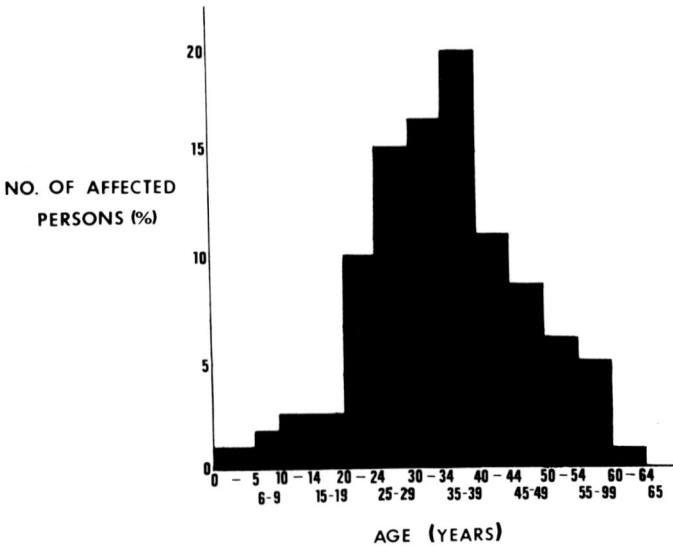

Fig.4.2. Range of ages at onset of 219 patients in the South African survey

Muncey (1916) reported a low age at onset (37.8) but also included a few juvenile patients, whereas Cameron and Venters (1967) documented a higher age at onset (43.2) and also a relatively high proportion of affected juveniles (4.8%). It would seem that there are influences other than the number of juvenile patients which may modify the mean age at onset in any given community.

Brackenridge (1973) has documented the lowest mean age at onset of Huntington's chorea of all reports. However, he based his findings on a review of the literature and the estimated result was very probably biased by the uneven reporting of different forms of the disease, with particular emphasis on those with precocious onset.

4.1.2. The Relevance of Age at Onset to Genetic Counselling

Even though some investigators (Caro 1977) have regarded "age at onset as a very vague retrospective point which is notoriously difficult to assess" and subsequently discarded this estimate from statistical analysis, the age at onset is of great importance and has major relevance to genetic counselling.

The genetic counsellor is commonly faced with the problem of a healthy person at risk for Huntington's chorea, who is concerned about the implications of the disease for himself and his children. As Huntington's chorea is an autosomal dominant disorder a child of an affected person has an even chance (50%) of having inherited the gene. The risk for each of his children in turn is 50% of 50%, i.e. 25%. However, the person in question has already passed through part of the risk period without being affected. The longer this continues, the more likely it is that he is homozygous for the normal allele and that he will remain unaffected. It is possible to give an actual risk estimate if the range of ages at onset of Huntington's chorea in that specific community is known. For example, if it is assumed that he was 49 years of age and living in Northamptonshire (Oliver, Table 4.2), by that age 80% of all heterozygotes would have already shown clinical signs of the disease. He is therefore either among the 20% of persons who have the abnormal gene and who manifest later in life, or belongs to the normal population in the sense that he does not have the gene for Huntington's chorea. By using conditional probabilities (Vogel and Motulsky 1979) the estimate of his risk of still developing Huntington's chorea is 16%, whilst that for his children is thus 8%. The details concerning this calculation are given in Appendix 1.

Stevens (1973) has drawn up a curve from which risk of heterozygosity at various ages can be read for offspring of affected patients in the area of Leeds, England (Fig. 4.3). A series of curves which take into account the ranges of parental ages at onset and their modifying effect on the risk to clinically normal offspring has also been constructed (Fig. 4.4). These curves offer rapid detection of the risk, which is a distinct advantage over the methods described previously.

Whilst the chances of inheriting the gene for Huntington's chorea are 50% for every child of an affected parent, this risk changes as the child gets older. A more accurate

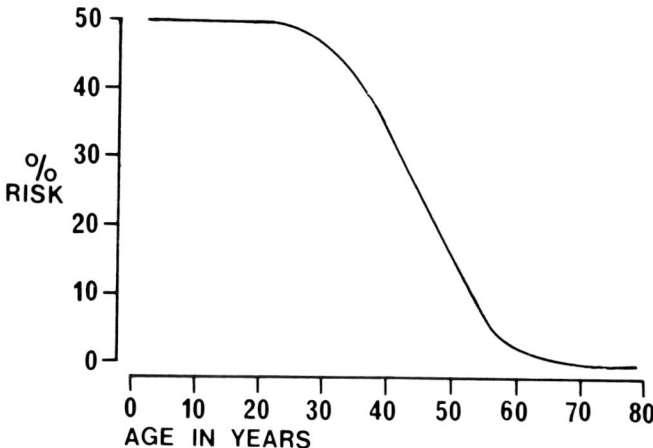

Fig.4.3. Curve from which risk of heterozygosity at various ages can be read for children of persons with Huntington's chorea. (From Stevens 1973.)

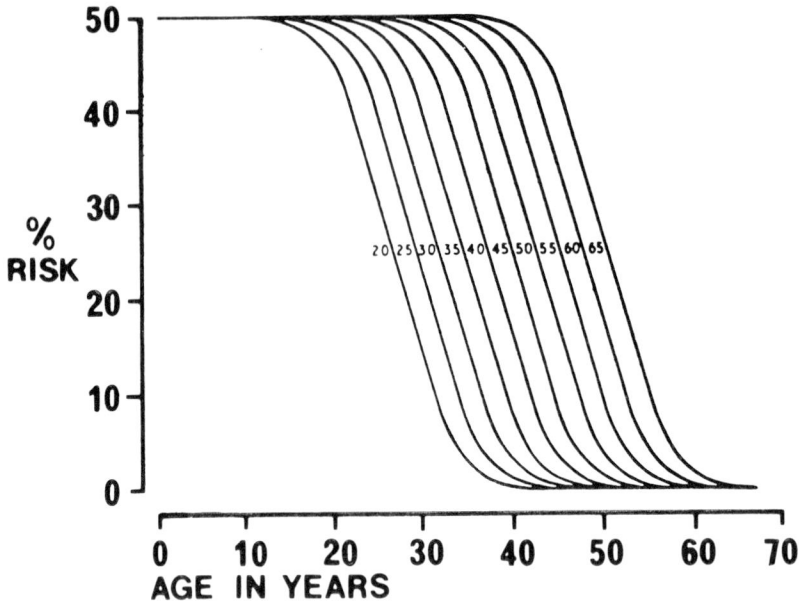

Fig.4.4. Curves from which risk of heterozygosity can be read, taking into account the age at onset of the disease in the affected parent. (From Stevens 1973.)

assessment of the genetic risk in the way described may enable the genetic counsellor to give some encouragement and more realistic advice to those who live with the long-term threat of developing Huntington's chorea.

4.2. Age at Death

The age at death of persons with Huntington's chorea is a specific point in time and is not subject to the same sources of potential bias as other measurements of the natural history of this disorder. A possible cause of distortion of the data may be premature death due to other factors, such as an accident, an unrelated intercurrent illness or suicide. In most investigations this has not been considered and thus the reported ages at death may be slight underestimates of the true figure. However, this possible source of error is minimised in a large population, in which ascertainment is near-complete.

4.2.1. International Comparison

The age at death from Huntington's chorea in different surveys is shown in Table 4.3. The reported results are fairly consistent, ranging from 51.4 (Brothers 1964) to 56.9 years (Stevens 1976). Those surveys showing low death ages (Bell 1934; Brackenridge, 1971; Brothers, 1964; Hayden, 1979; Panse 1942; Reed et al. 1958) also have low ages at onset. The higher ages at death recorded in some investigations (Bolt 1970; Stevens 1976) parallel the raised mean age at onset of the disease. This would suggest that there is little variation in duration of the disease in these different studies. A higher proportion of juvenile patients, a major factor in the lowering of the mean age at onset, also results in the depression of the mean age at death.

Table 4.3. Age at death (AaD): previous reports

Location	Author(s)	Year	No. of patients	Mean AaD
Literature review	Bell	1934	349	53.27
Rhineland	Panse	1942	473	52.2
Switzerland	Zolliker	1949	143	54.5
Norway	Ørbeck and Quelprud	1954	117	55.5
Michigan, USA	Reed et al.	1958	262	53.58
Germany	Wendt	1960	246	56.37
Victoria, Australia	Brothers	1964	123	51.4
Scotland	Cameron and Venters	1967	194	54.6
Belgium	Husquinet	1969	231	54.1
France	Petit	1969	141	55.6
Scotland	Bolt	1970	269	56.7
Literature review	Brackenridge	1971	403	51.7
Minnesota, USA	Marx	1971	318	54.1
Leeds, England	Stevens	1976	256	56.9
East Anglia	Caro	1977	500	55.9
South Africa	Hayden	1979	160	51.87

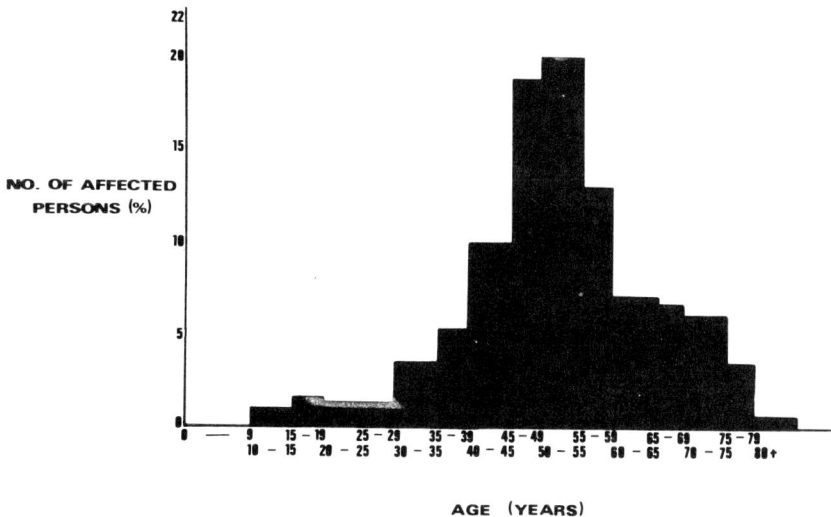

Fig.4.5. The range of ages at death of 160 patients with Huntington's chorea in South Africa

In general, the majority of investigations show no sex difference in ages at death. The range of ages at death for 160 affected individuals in a large survey is shown in Fig. 4.5. It can be seen that whilst most affected persons die between the ages of 50 and 60 years, death from Huntington's chorea can also occasionally occur at the extremes of life.

When considering ages at death due to Huntington's chorea in a given community it is important to consider the overall health and life expectancy in that region. The lower mean age at death from Huntington's chorea may reflect a reduced life expectancy for all persons in that particular country. This influence may be of special significance in less sophisticated areas where lack of medical care may result in premature death from infection, malnutrition and trauma.

Table 4.4. Duration of Huntington's chorea: previous reports

Author(s)	Location	Year	No. of patients	Mean duration (years)
Bell	England	1934	204	13.7
Panse	Rhineland	1942	446	13.4
Zolliker	Switzerland	1949	202	14.0
Reed et al.	Michigan, USA	1958	153	15.8
Brothers	Victoria, Australia	1964	97	12.2
Cameron and Venters	Scotland	1967	127	10.6
Petit	France	1969	65	14.4
Bolt	Scotland	1970	176	14.6
Brackenridge	Australia	1971	191	11.9
Stevens	Leeds, England	1976	180	13.6
Hayden	South Africa	1979	94	14.1

4.3. Duration

Since the duration of Huntington's chorea is derived by subtracting the age at onset from the age at death, it is subject to the same potential sources of bias as these other parameters.

Various estimates of the duration of the disease are shown in Table 4.4. There is no significant geographical or sex-related variation in this parameter in various parts of the world. An increased duration of Huntington's chorea in affected females compared with affected males has been recorded on two occasions (Dewhurst et al. 1970; Stevens 1976) but these findings have not been substantiated. The duration of Huntington's chorea and age at death have remained essentially constant over the past 50 years, thus reflecting the failure of medical therapy to prolong affected persons' lives.

The range of the duration of the disorder in patients in South Africa is shown in Fig. 4.6, and it is evident that the vast majority of patients suffer from the disease for 10 to 20 years. The disease spanned less than 5 years only in those instances where death occurred from other, unrelated causes such as accidents or suicide. Affected juveniles have a significantly shorter duration of the illness than those with adult onset. On further separation of the juvenile group into those with childhood and those with adolescent onset, it is evident that the mean duration for those with onset before the age of 10 is particularly short (Table 4.5.). Persons with later age at onset (over 60 years) may also have a shorter duration; this is mainly a reflection of the reduction in life expectancy in people at this age. It is evident, therefore, that age at onset may be a valuable clue to estimation of the duration of the disorder, particularly when the disease presents at the extremes of life (Fig. 4.7).

Table 4.5. Comparison of duration in adult and juvenile patients

	No. of patients	Mean duration (yr)	s.e.m.	Significance
Adults:	86	14.65	0.49	$P < 0.001$
Juveniles:	8	9	0.46	

From Hayden (1979).

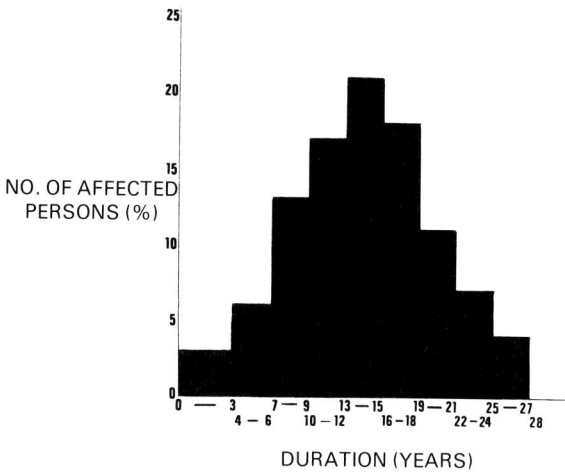

Fig.4.6. The varying durations of Huntington's chorea in patients in South Africa

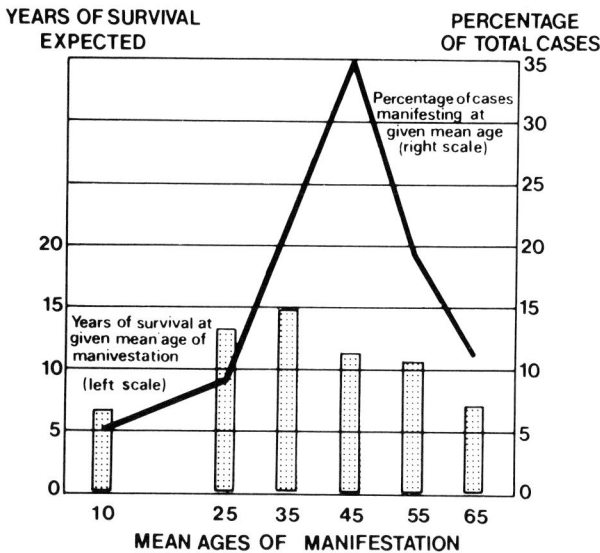

Fig.4.7. The relationship between age at onset and duration of Huntington's chorea. (From Scoones 1980)

4.4. Factors Modifying the Action of the Gene

The geographical differences in the parameters concerning the natural history of Huntington's chorea can partly be attributed to variations in techniques for the collection of data. However, there are genetic and environmental factors that may modify the action of the gene for Huntington's chorea and which must be considered in this context.

Knowledge of those factors which hasten onset of the disease would be important for two major reasons. Firstly manipulation of known environmental factors may delay its onset. Secondly this knowledge would be of great potential use in genetic counselling and allow more accurate prediction of the age at onset and course of the disease in any person, taking into account their particular genetic and environmental background.

4.4.1. Genetic Considerations

Several investigators have addressed themselves to the problem of determining whether the variability in the natural history and clinical presentations of Huntington's chorea in different individuals is genetically determined. The parameter most frequently examined has been age at onset, and analysis of variance which compares the variation within and between families is a most convenient method for testing for the genetic contribution.

Results of different studies (Reed et al. 1958; Mattsson 1974; Stevens 1976) consistently show that within-kindred variation of age at onset is less than that between kindreds. Reed et al. (1958) have suggested that this is probably due to a greater genetic similarity within than between families. Whether this is because of the possession of similar 'background genes' or is the result of members of the same family inheriting common modifying genes is uncertain. In an alternative hypothesis Wallace (1972) has proposed heterogeneity as the reason for this phenomenon.

Twin data are particularly useful in this situation. The twin method is based on the fact that monozygotic twins originate from the division of one zygote and must therefore be genetically identical. If it is required to assess whether a character is genetically determined, the degree of similarity between monozygotic twins for a given parameter can

Table 4.6. Reports of twins with Huntington's chorea

Author(s)	Age at onset A	B	Sex	Comment[a]
Russell (1894)	27	27	M	Identical physically
Rosenthal (1927)	29	?	M	Twin B unaffected at 40
Rosanoff and Handy (1935)	35	35	F	Identical physically
Entres (1940)	41	41	F	Identical
	45	45	M	Identical
Jequier (1945)	44	44	M	Identical
Jelgersma (1957)	45	?	F	Probably identical
Parker (1958)	45	45	M	Identical
Myrianthopoulos and Rowley (1960)	22	22	F	MZ proven
Perrine and Goodman (1966)	26	36	M,F	Dizygotic. Discordant clinically
Schioltz-Christensen (1969)	24	24	F	MZ proven
Bolt (1970)	?	?	F	Details incomplete
Oepen (1973)	22	25	F	Discordant clinically. MZ proven
Husquinet et al. (1973)	21	22	F	MZ proven
Bird and Omenn (1975)	±26	±28	F	Discordant clinically. MZ proven
Bachman et al. (1977)	20	20	F	MZ proven

[a]MZ, monozygosity.

be measured. Dizygotic twins develop from two separate eggs and are no more similar genetically than normal sibs.

There have been numerous reports of affected twins with Huntington's chorea (Table 4.6). However, systematic twin studies in which monozygosity has been clearly established have been performed in only a few instances. The determination of zygosity on the basis of physical similarity is not always easy, and where possible this can best be confirmed by means of serologic studies. Of the six proven sets of monozygotic twins, three had onset of the disease in the same year, whilst the remaining three had first symptoms of Huntington's chorea within 3 years of each other. Another six sets of twins were thought to be monozygotic by virtue of their marked physical similarity. Both members of each of these sets of twins had the same ages at onset. Taken together all the twins thought to be monozygotic had ages of onset which did not differ by more than 3 years from each other. This may be preliminary evidence for a major genetic contribution to the determination of this parameter in Huntington's chorea.

Variation in clinical features between sibs of a particular sibship is well recognised, but it is difficult to establish whether these differences are genetic or environmental in origin. It is noteworthy that in two of the later reports (Bird and Omenn 1975; Oepen 1973) where monozygosity was proven, individual twins showed a marked difference in clinical expression of the disease. In Oepen's publication (1973) one of the sisters has the rigid form of the disease, in contrast to the typical choreiform movements of her sib. A disparity in the extent of chorea is also seen in the twins mentioned in Bird and Omenn's report. The divergent neurological findings in these two sets of monozygotic twins suggest that the clinical presentation is only partly influenced by the affected person's genetic constitution. Non-genetic factors are clearly also important in this respect.

Formal twin-method studies will yield important information which may clarify the relative effects of genes and the environment in the determination of the different features of Huntington's chorea.

4.4.2. Environmental Factors

There is some evidence that environmental factors influence the age at onset of Huntington's chorea. Patients often ascribe onset of the disease to some major happening in their lives, such as pregnancy (Spengler 1956), trauma and infection (Korenyi et al. 1972).

Stevens (1976) commented that there is a general tendency for the mean age at onset to be higher in Europe than in non-European countries such as America, South Africa and Australia. Whether these geographical differences are significant is difficult to assess in view of the problems inherent in the definition of age at onset.

Brackenridge (1974) has analysed data from the literature and his findings may partly explain this phenomenon. He showed that there is a statistically significant decrease in age at onset as environmental temperature rises. The higher mean annual temperature in South Africa, Australia and some parts of North America, compared with Europe, parallels the lower mean age at onset of Huntington's chorea in these areas. Whilst this concept is attractive in theory, the means by which this effect could be mediated are at present unknown. Genetic factors can probably be discarded as a reason for this finding, as the immigrants to the USA, South Africa and Australia are largely of north-west European origin and thus have the same genetic composition.

At the present time knowledge concerning the effect of the environment on the action of the gene for Huntington's chorea is scanty and from the data presented no definite conclusions can be drawn.

References

Bachman D, Butler I, McKhann G (1977) The long term treatment of juvenile Huntington's chorea with dipropylacetic acid. Neurology (Minncap) 27: 193-197

Bell J (1934) Huntington's chorea. In: Fisher RA (ed) The treasury of human inheritance, vol IV/1. Cambridge University Press, London, pp 1-77

Bickford JAR, Elisson RM (1953) High incidence of Huntington's chorea in the Duchy of Cornwall. J Ment Sci 99: 291-294

Bird TD, Omenn GS (1975) Monozygotic twins with Huntington's chorea in a family expressing the rigid variant. Neurology 25: 1126-1129

Bolt JM (1970) Huntington's chorea in the West of Scotland. Br J Psychiatry 116: 259-270

Brackenridge CJ (1971) The relation of type of initial symptoms and line of transmission to ages at onset and death in Huntington's disease. Clin Genet 2: 287-297

Brackenridge CJ (1973) Interrelations between some clinical factors in Huntington's chorea. In: Barbeau A, Chase TN, Paulson GW (eds) Huntington's chorea, 1872-1972. Raven Press, New York, pp 65-73

Brackenridge CJ (1974) Effect of climatic temperature on the age of onset of Huntington's chorea. J Neurol Neurosurg Psychiatry 37: 297-301

Brothers RCD (1964) Huntington's chorea in Victoria and Tasmania. J Neurol Sci 1: 405-420

Cameron D, Venters GA (1967) Some problems in Huntington's chorea. Scott Med J 12: 152-156

Caro A (1977) Huntington's chorea: a genetic problem in East Anglia. PhD thesis, University of East Anglia

Davenport CB, Muncey EB (1916) Huntington's chorea in relation to heredity and eugenics. Am J Insan 73: 195-222

Dewhurst K, Oliver JE, Mcknight AL (1970) Socio-psychiatric consequences of Huntington's disease. Br J Psychiatry 116: 255-258

Entres JL (1921) Zur Klinik und Vererbung der Huntingtonschen Chorea. Monographien aus dem Gesamtgebiete der Neurologie und Psychiatrie, Heft 27. Springer, Berlin

Entres JL (1940) Der Erbvietstanz Huntingtonsche Chorea: erbbiologischer Teil. In: Gutt A (ed) Handbuch der Erbkrankheiten, vol III. Thieme, Leipzig, pp 243-262

Hayden MR (1979) Huntington's chorea in South Africa. PhD thesis, University of Cape Town

Heathfield KWG (1967) Huntington's chorea: investigation into prevalence of this disease in the area covered by the North East Metropolitan Hospital Board. Brain 90: 203-232

Husquinet H (1969) La chorée de Huntington dans quatre provinces belges. In: Rapports 67e session, Congrès de Psychiatrie et Neurologie de Langue Française. Masson et Cie, Paris

Husquinet H, Frank G, Vrankx C (1973) Detection of future cases of Huntington's chorea by the L-dopa load test: Experiment with 2 monozygotic twins. In: Barbeau A, Chase TN, Paulson GW (eds) Huntington's chorea, 1872-1972. Raven Press, New York, pp 301-310

Jelgersma HC (1957) Monozygotic twins with concordant Huntington's chorea and discordant hemiplegia. Folio. Psychiatr Neerl 60: 50-52

Jequier M (1945) La chorée de Huntington. Arch Julius Klaus Stift 20: 77-208

Korenyi C, Whittier JR, Conchado D (1972) Stress in Huntington's disease. Dis Nerv Syst 33: 339-344

Lyon RL (1962) Huntington's chorea in the Moray Firth area. Br Med J i: 1301-1306

Marx R (1971) Huntington's disease in Minnesota. PhD thesis, University of Minnesota

Mattsson B (1974) Clinical, genetic and pharmacological studies in Huntington's chorea. UMEA University Medical Dissertations 7. UMEA, Sweden, pp 21-51

Minski L, Guttmann E (1938) Huntington's chorea: a study of 34 families. J Ment Sci 84: 21-96

Myrianthopoulos N, Rowley P (1960) Monozygotic twins concordant for Huntington's disease. Neurology 10: 506-511

Oepen H (1973) Discordant features of monozygotic twin sisters with Huntington's chorea. In: Barbeau A, Chase TN, Paulson GW (eds) Huntington's chorea, 1872-1972. Raven Press, New York, pp 199-201

Oliver JE (1970) Huntington's chorea in Northamptonshire. Br J Psychiatry 166: 241-253

Ørbeck AL, Quelprud T (1954) Setesdalsrykka (chorea progressiva hereditaria). Nor Videnskap Acad Schr 1: 1-125

Panse F (1942) Die Erbchorea: eine klinische-genetische Studie. Samml Psychiatr Neurol Einzeldarst 18. Thieme, Leipzig

Parker N (1958) Observations on Huntington's chorea based on a Queensland survey. Med J Aust. I: 251-259

Perrine GA, Goodman RM (1966) A family study of Huntington's chorea with unusual manifestations. Ann Intern Med 64: 507-574

Petit H (1969) Rapports 67e session, Congrès de Psychiatrie et Neurologie de Langue Française, Brussels. Masson, Paris

Reed TE, Chandler JH, Hughes EM, Davidson RT (1958) Huntington's chorea in Michigan: demography and genetics. Am J Hum Genet 10: 201-225

Rosanoff A, Handy L (1935) Huntington's chorea in twins. Arch Neurol 33: 839-841

Rosenthal C (1927) Zur Symptomatologie und früh-Diagnostik der Huntingtonsche Krankheit. Z Gesamte Neurol Psychiatr 3: 254-269

Russell JW (1894) Two cases of hereditary chorea ocurring in twins. Birmingham Med Rev 35: 31-33

Schioltz-Christensen E (1969) Chorea Huntington and epilepsy in monozygotic twins. Eur Neurol 2: 250-255

Scoones T (1980) Huntington's chorea. Office of Health Economics Publication 67. OHE, London, p7

Shokeir MH (1975) Investigations on Huntington's disease in the Canadian prairies. II. Fecundity and fitness. Clin Genet 7: 349-353

Sjogren T (1936) Vererbungsmedizinische Untersuchungen über Huntingtons Chorea in einer schwedischen Bauernpopulation. Z Menschl Vererb-Konstit-Lehre 19: 131-165

Spengler M (1956) Verschlimmerung von Chorea Huntington in der Gravidität. Schweiz Med Wochenschr 86: 931-932

Spillane J, Phillips R (1937) Huntington's chorea in South Wales. Q J Med 6: 403-423

Stevens DL (1973) The heterozygote frequency for Huntington's chorea. In: Barbeau A, Chase TN, Paulson GW (eds) Huntington's chorea, 1872-1972. Raven Press, New York, pp 191-198

Stevens DL (1976) Huntington's chorea. a demographic, genetic and clinical study. MD thesis, University of London, pp 1-338

Vogel F, Motulsky A (1979) In: Human genetics: problems and approaches. Springer, Berlin, Heidelberg, New York, pp 595-596

Wallace DC (1972) Huntington's chorea in Queensland: a not uncommon disease. Med J Aust 2: 1275-1277

Wendt GG (1960) Praktische Erfahrungen bei der Sammlung aller Falle von Huntingtonscher Chorea. In: Die spontane und induzierte Mutationsrate bei Versuchstieren und beim Menschen, Strahlenschutz Vol 17. Gersbach, München, pp 1-36

Wendt GG, Drohm D (1972) Fortschritte der allgemeinen und klinischen Humangenetik: vol IV, Die Huntingtonsche Chorea. Thieme, Stuttgart

Wendt GG, Londzettel I, Unterreiner I (1959) Erkrankungsalter bei der Huntingtonscher Chorea. Acta Genet (Basel) 9: 18-32

Zolliker A (1949) Die Chorea Huntington in der Schweiz. Schweiz Arch Neurol Psychiatr 64: 448-457

5 Clinical Features

The diagnosis of Huntington's chorea depends on establishing a positive family history for the disorder and on a detailed clinical assessment. Whilst the cardinal clinical features are chorea and dementia, some patients may present with a wide range of other signs and symptoms. This broad spectrum of clinical manifestations may be confusing and result in misdiagnosis.

George Huntington (1872) failed to appreciate that there are major variations from the classic clinical picture. This misunderstanding has persisted to the present day and is partly due to the fact that very little attention has been given to a detailed analysis of the clinical features of Huntington's chorea. Recent comprehensive reviews of this subject are to be found in the *Handbook of Clinical Neurology* by Bruyn (1968), in the clinical section of Stevens' thesis (1976) and in Shoulson's report to the Commission for the Control of Huntington's Disease (1977).

In this chapter the outstanding clinical characteristics, the proposed staging system and the different variants of Huntington's chorea will be reviewed. This will be followed by a discussion of the problems of diagnosis and different diagnostic techniques.

5.1. The Presenting Symptoms and Signs

The majority of investigators have reported that neurological features comprise the predominant initial symptom (Table 5.1). In almost every instance, however, close questioning of the spouse of a patient reveals that the disease is ushered in by subtle mental changes. These alterations most often take the form of a change in personality with irritability, moodiness, irascibility and impulsiveness as the predominant features. Depression may also be amongst these early symptoms and in these instances spouses often complain of a general poverty of affect and ideation.

Although the signs at this early stage are minimal, there may be obvious disturbance of functional ability. The housewife may find that she is less able to look after the home and the businessman may find it more difficult to maintain his appointments and manage his financial affairs. These changes are difficult to define and the definitive diagnosis is usually only made when there are more easily discernible signs.

The single most common complaint is jerkiness, clumsiness or mild incoordination, which usually heralds the onset of chorea. Whilst these findings may not be present at rest, certain manoeuvres can be employed to elicit these abnormalities. Persons thought to have the disease can be asked to squeeze two of the examiner's fingers and maintain a constant pressure for 5 s. Spontaneous choreiform movements will often be evoked, producing the characteristic 'milkmaid's sign' (Fig 5.1). At this early stage affected

Table 5.1. Reported percentage of survey population presenting with neurological or psychiatric features

Author(s)	Population size	Neurological	Psychiatric	Combined
Oepen (1963)	219	34	24	42
Brothers (1964)	237	59	27	14
Heathfield (1967)	84	54	46	—
Bolt (1970)	68	54	27	19
Brackenridge (1971)	292	39	28.8	32.2
Mattsson (1974)	162	22	48	30
Stevens (1976)	102	65	35	—
Hayden (1979)	130	39	51	10

Fig.5.1. Spontaneous choreiform movements can be elicited in persons with Huntington's chorea on squeezing the examiner's fingers — the 'milkmaid's sign'

individuals may also be unable to perform complex facial movements, such as whistling, blowing out the cheeks, frowning, putting out the tongue or showing the teeth (Fig. 5.2). This could be viewed as a form of facial apraxia and may be elicited before the onset of obvious chorea. If a patient is asked to walk on a straight line this may also evoke abnormal, irregular swaying movements of the whole body which are not present at rest.

A small proportion of affected persons never develop the typical choreiform movements and may present with a general slowing of all movements associated with a progressive rigidity. In a few instances patients may have no symptoms referable to the basal ganglia but have defects in judgement, affect and memory, whilst others may present with a true

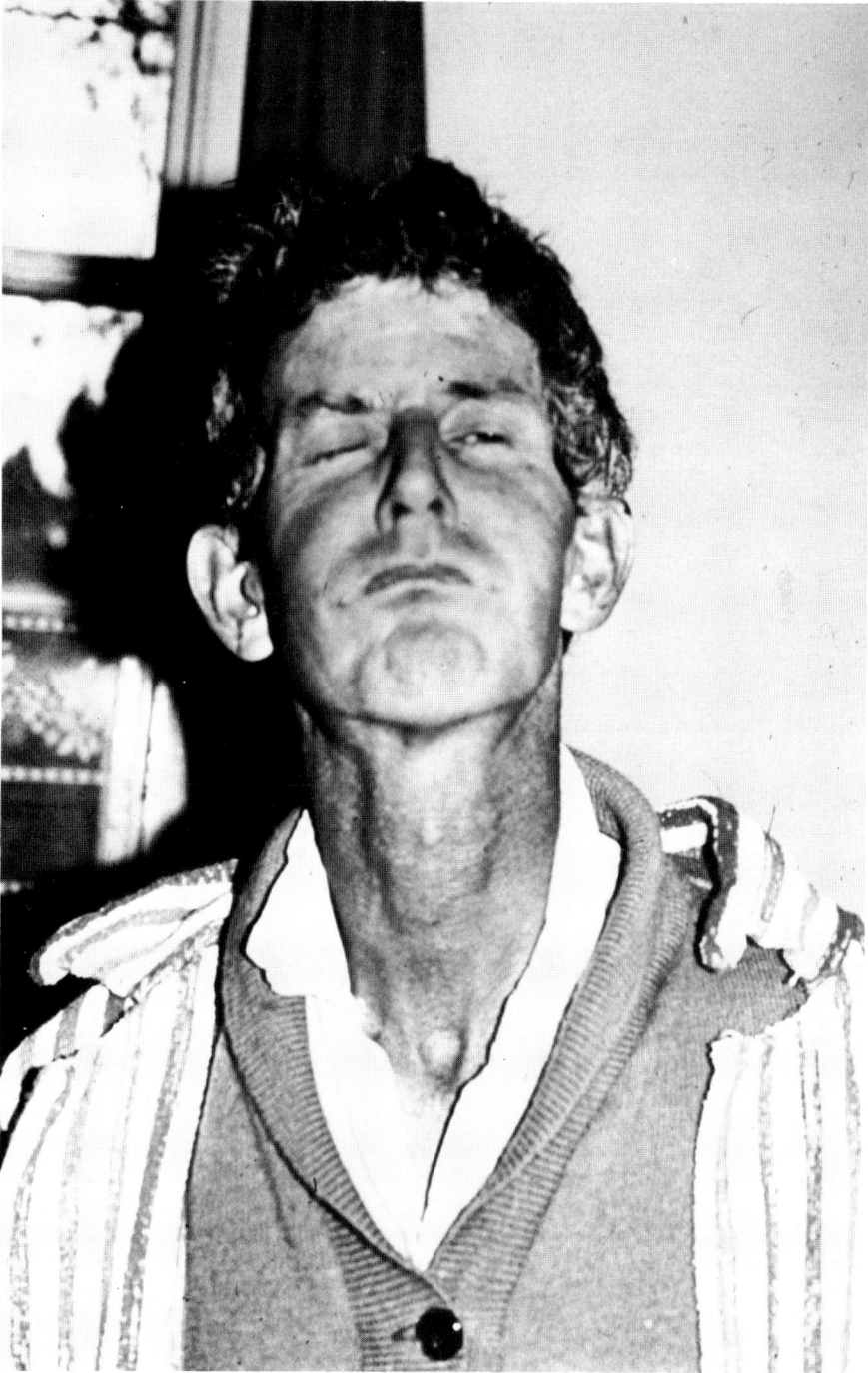

Fig.5.2. The inability to perform sustained complex facial movements is a common early sign

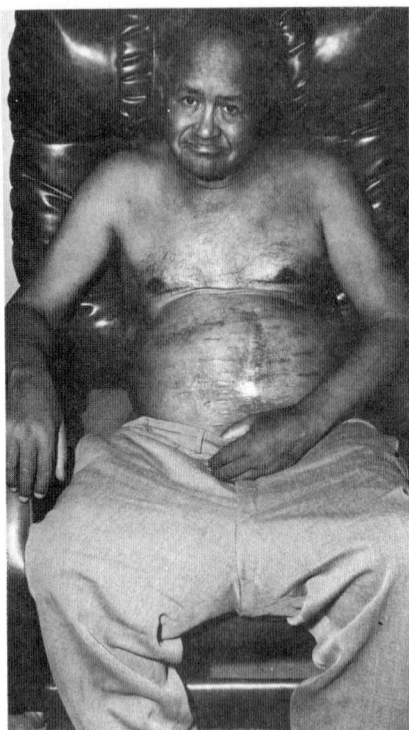

Fig.5.3. This patient with definitively diagnosed Huntington's chorea is obese. His predominant motor abnormality is chorea

psychosis with auditory and visual hallucinations and paranoid delusions. These signs and symptoms may, understandably but wrongly, be attributed to a schizophrenic psychosis. The involuntary movements may only appear some years later.

Patients who present with chorea are more likely to be diagnosed early and correctly, whilst other persons whose initial symptomatology is psychiatric are more likely to be initially misdiagnosed. The importance of these facts is to reiterate that Huntington's chorea may have a wide variety of subtle clinical presentations. Awareness of this phenomenon should lead to earlier diagnosis, which will facilitate more appropriate genetic counselling.

5.2. General Manifestations

A striking feature in many patients is severe emaciation in spite of excellent appetites. Bruyn (1968) commented that none of the 100 or more patients whom he had examined were obese. Whilst generalised weight loss frequently occurs, nevertheless a few patients with Huntington's chorea may, in fact, be overweight (Fig. 5.3).

Whittier (1967) has suggested that the "catabolism exceeds that which might be expected from the energy demand generated by chorea or rigidity". In this context there is some evidence that the marked weight loss may be a manifestation of the underlying biochemical defect and may not purely reflect an imbalance between food intake and energy expenditure (Sandburg and Fibiger 1979). Thyroid function has been investigated in this regard but no abnormalities have been reported (Hayden and Vinik 1979).

Fig.5.4. The picture between this man's legs was taken on his 50th birthday, 1 year before this photograph was shot. The rapid deterioration in his condition is easily discernible

Almost all adult patients look older than their chronological age. Premature greying and loss of skin turgor, with early wrinkling are common. The appearance of an affected person 5 years after the onset of his disease and 1 year after the photograph placed between his legs was taken, is shown in Fig. 5.4. The ageing effect in this man is easily discernible.

A few patients may complain of excessive sweating. Dynan (1913-1914) and Refsum (1938) have previously commented on this feature. Aminoff and Gross (1974) have investigated vasoregulatory activity in Huntington's chorea and found a normal resting blood pressure but a significantly greater fall in blood pressure on tilting than that seen in control subjects. There has been no study which has confirmed these findings and at present this is the only firm evidence for autonomic dysfunction in Huntington's chorea.

Many unrelated patients bear a remarkable similarity to each other; this is not surprising, in view of the similar pathological process due to the common genetic defect.

5.3. Neurological Features

5.3.1. Chorea

The most striking clinical feature in this disorder is chorea, which is obviously apparent in approximately 90% of all affected individuals. Initially the chorea may be minimal, but as the disease progresses the patient will present with ceaseless writhing, jerking and twisting of different parts of the body as a result of the interruption of the voluntary by

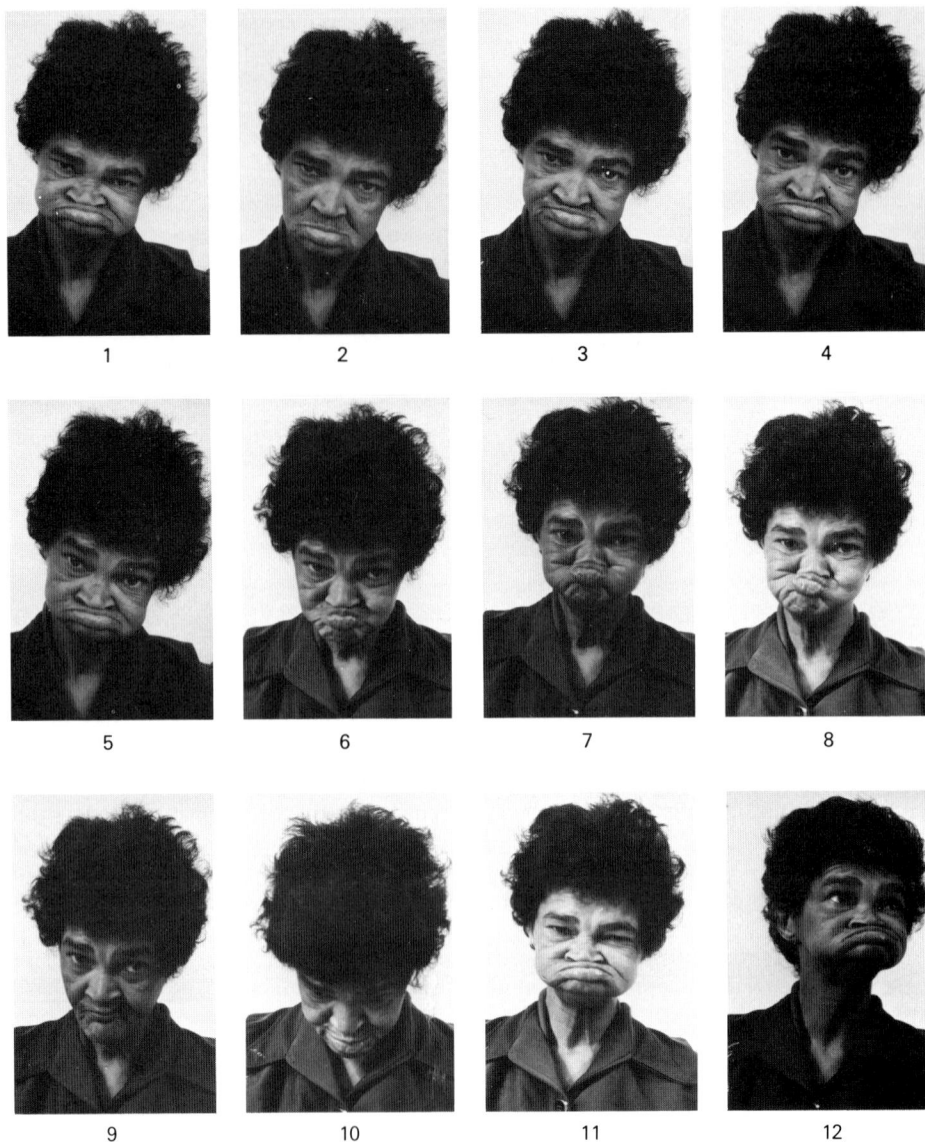

Fig.5.5. These sequential photographs were taken at 1-s intervals, and show gross choreiform movements of the face and neck

involuntary movements. In some persons chorea may mainly affect the face, whilst in others the extremities will predominantly exhibit the hyperkinesia.

Involuntary facial gestures are common and present a fairly typical appearance. Marked pouting of the lips together with twitching of the cheeks and irregular elevation of the eyebrows is a frequent sequence (Fig. 5.5). Intermittent protrusion of the tongue may also be evident. In some persons these facial movements are associated with massive excursions of the head, which follows a complete rotatory sweep on its axis as first it is jerked backwards and then bowed. Whilst the pattern of the movements is different between affected persons, they seem to recur in each patient in a stereotyped manner.

When sitting at ease many patients show constant irregular motion of the hands and legs. The legs may be alternately crossed and uncrossed. Fingers are constantly flexed and extended and these movements are exaggerated when the patient holds his hands before him (Fig. 5.6). Kinnier Wilson (1954) has described the typical 'choreic hand', in which the wrist is often flexed with hyperextension of the metacarpophalangeal joints and the fingers themselves are held straight and apart. This type of hand posture may be seen in patients with chorea of varied causation, including Huntington's chorea and Sydenham's chorea. However, in contrast to the brusque, abrupt and jerky motions of the latter condition, movements in Huntington's chorea are generally slower and more athetoid in nature. Whilst chorea is the characteristic feature of this disorder, other abnormal movements such as tremor and intention myoclonus may also occasionally be present (Novom et al. 1976).

The hyperkinesia of Huntington's chorea may show major variations in severity in each patient. Emotion, excitement and tension exacerbate the symptoms whilst the chorea ceases during sleep. Bruyn (1973) has noted that the chorea becomes less conspicuous and more athetoid as the disease progresses, a feature previously commented on by Denny-Brown (1962). This may in part be due to the advent of pyramidal tract dysfunction.

A small proportion of patients ($\pm 5\%$) will not show any involuntary movements at any stage of their illness and in these individuals rigidity is often the predominant feature.

5.3.2. Hypertonicity: Rigidity and Spasticity

Approximately 45% of patients with Huntington's chorea have evidence of increased tone. This may either be extrapyramidal (lead-pipe or cogwheel) or pyramidal (clasp-knife) in origin and occurs in three different clinical situations:

Group A: Rigidity *ab initio* without chorea ($\pm 5\%$ of patients).

Group B: Rigidity and/or spasticity occurring simultaneously with chorea ($\pm 20\%$ of patients).

Group C: Choreiform movements being superseded by spasticity, usually in association with other signs of pyramidal tract dysfunction ($\pm 20\%$ of patients).

Whilst it is occasionally difficult to differentiate between extrapyramidal and cortical (pyramidal) hypertonicity, in most instances this is possible from consideration of the nature of the increased tone and the associated signs. It is noteworthy that cogwheel rigidity, spasticity and chorea may on occasions be elicited simultaneously in a single patient.

Signs suggestive of an extrapyramidal cause for hypertonicity include a lead-pipe or cogwheel type of rigidity, a tremor, a mask-like facies (Fig. 5.7), a positive glabellar reflex, bradykinesia, lack of associated movements in walking and a flexed posture. Hamilton in

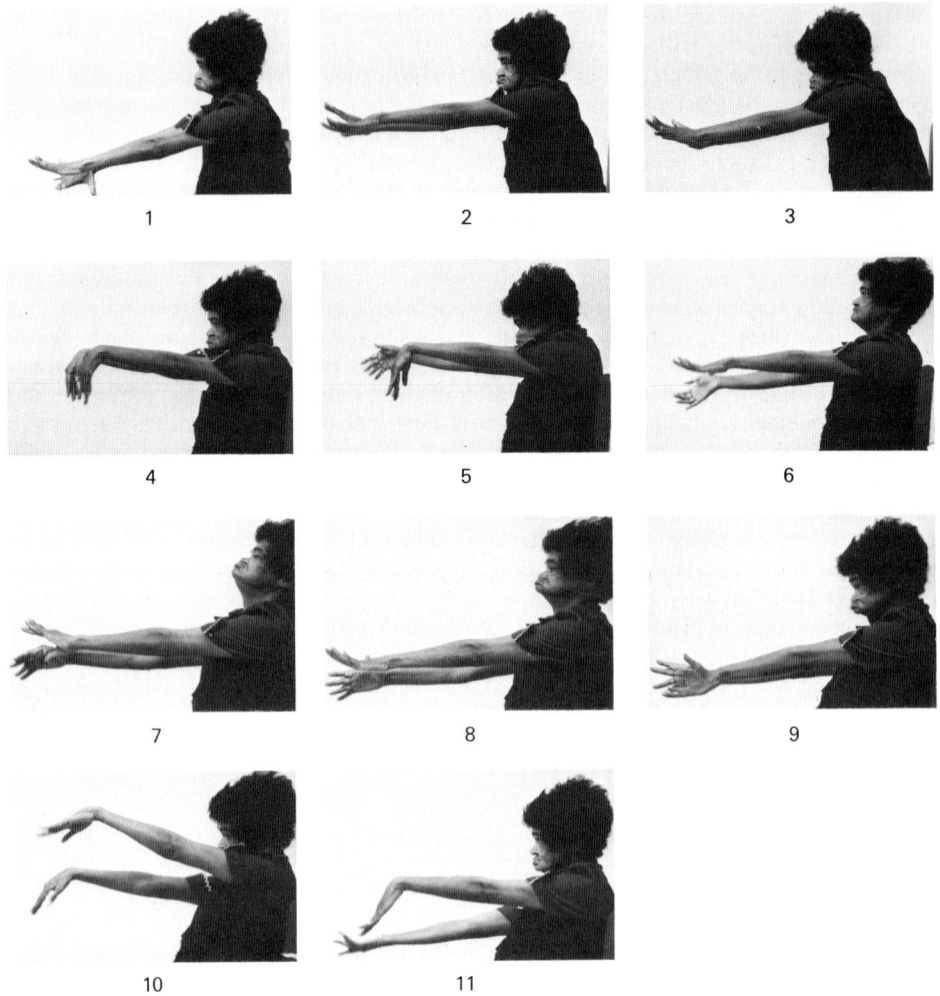

Fig. 5.6. On extension of the arms the choreiform movements are exaggerated. (Photographs were taken at 1-s intervals)

America (1908) was one of the first to document the Parkinsonian-like features seen in Huntington's chorea. The two most comprehensive studies in this regard are those of Bittenbender and Quadfasel (1962) who, in addition to their own eight patients, reviewed a further 62 from the world literature, and Bruyn (1968), who summarised the findings in over 100 similarly affected patients.

Rigidity occurs more commonly at the extremes of the age spectrum. The hypertonicity in juvenile patients will be discussed in Sect. 5.7. Approximately a third of all patients develop increased tone in association with other signs of pyramidal tract dysfunction as the disease progresses. A detailed description of these features is given in Table 5.2. Many of these different signs are commonly seen simultaneously in one patient. The signs listed are indicators of cortical atrophy and/or damage to the pyramidal tracts.

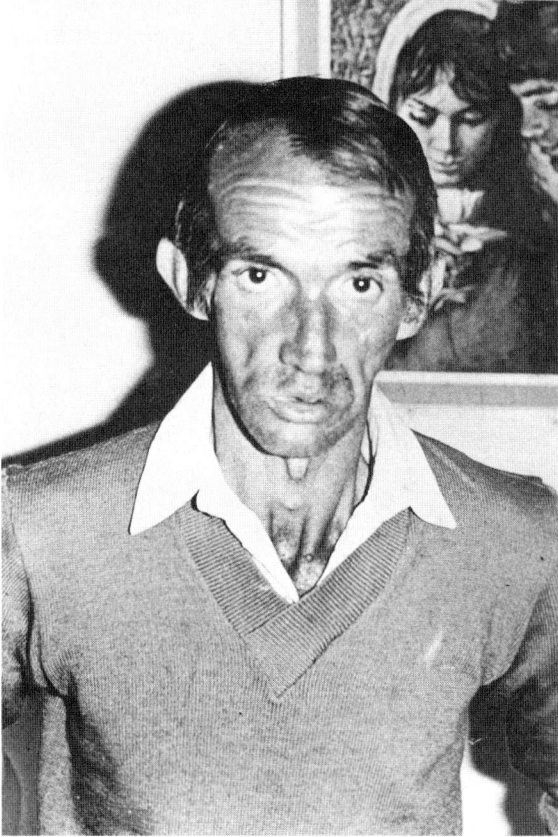

Fig.5.7. This man has a mask-like facies which, together with a positive glabellar reflex and lead-pipe rigidity, is reminiscent of Parkinsonism

Table 5.2. Frequency of signs of pyramidal tract dysfunction

Signs	Frequency (%)
Hypertonicity (clasp-knife)	10-20
Hyperreflexia	26-64
Extensor plantars response	7-10
Absent abdominal reflexes	17-25
Clonus	13-20

Based on reports by Bittenbender and Quadfasel (1962), Heathfield (1967), Bruyn (1968), Stevens (1976) and Hayden (1979).

Bittenbender and Quadfasel (1962) concluded their article by asserting that the "frequency of occurrence of the rigid form of Huntington's chorea is greater than is generally appreciated". This statement probably is still valid today.

5.3.3. Dysarthria

All patients have some difficulty in articulation. A severe dysarthria may mistakenly lead the examiner to assume that the patient is grossly demented, when the problem is more that of expression than comprehension. The intellect may be only mildly impaired at this stage.

The dysarthria presents as a slowness or hesitancy in speech and later is characterised by gross disorganisation. The presence or severity of dysarthria shows no correlation with other clinical manifestations such as chorea or rigidity, and usually occurs early. This symptom results from an incoordination of the muscles of articulation and probably reflects the combined effects of pyramidal and extrapyramidal damage.

5.3.4. Dysphagia

Dysphagia is common in the terminal phases of Huntington's chorea. Most patients experience this problem and in many instances it results in aspiration of ingested food and subsequent bronchopneumonia, which may lead to death. Edmonds (1966) reported that 85% of affected persons in his series died from a respiratory cause, probably secondary to aspiration of fluids or solids.

The dysphagia is in part consequent to involuntary movements of the muscles of deglutition. Another possible contributing factor is bilateral damage to the pyramidal tracts, resulting in a pseudobulbar palsy. It is apparent that the latter influence is not great, as the symptoms of dysphagia and signs of other pyramidal tract dysfunction show no significant positive correlation.

Dysphagia occurs in both rigid and choreic patients. This is graphically shown in Fig. 5.8, which depicts two sisters with markedly different clinical features of the disease. The sister with predominant chorea has finished drinking long before her rigid sib.

5.3.5. Disturbance of Gait

The gait of patients with Huntington's chorea constitutes a characteristic feature of the disease. At rest most affected persons stand on a broad base and involuntary movements are usually contained. However, on attempting to walk the unsteadiness increases and is characterised by swaying jerkiness and precipitate advances during which the patient may fall. In the event of a loss of consciousness or headache after such an accident, the possibility of a subdural haematoma should be considered.

The movements are characterised by large lateral deviations from the straight line with displacement from side to side, resulting in what Heathfield (1967) has termed 'zig-zag' progression (Fig. 5.9). This has understandably resulted in numerous patients being presumed to be drunk and, in some instances, arrested for this supposed misdemeanour. Persons with rigidity as the predominant feature have a different gait, with slowness in initiating movements, retarded progression and a bent posture (Fig. 5.10). Both persons with chorea and those with rigidity have difficulty in turning around.

Associated movements whilst walking are also abnormal, the arms being kept rigidly at the sides or held in abduction to accommodate the choreiform movements of the trunk. Fractures, bruises and other injuries are commonplace and in many instances patients are unable to walk without assistance.

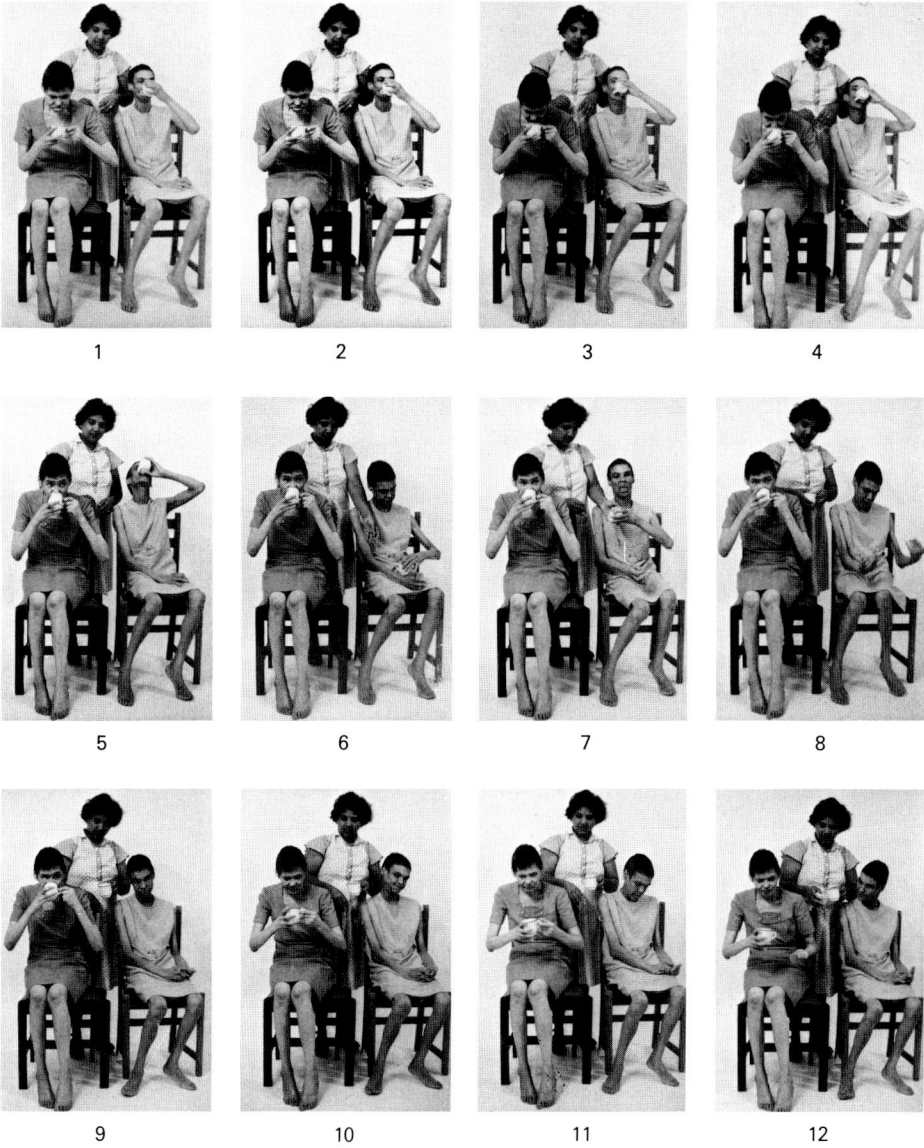

Fig. 5.8. The sister on the left, with predominant rigidity, takes a much longer time to finish the cup of water than her choreic sister. (Photographs were taken at 2-s intervals)

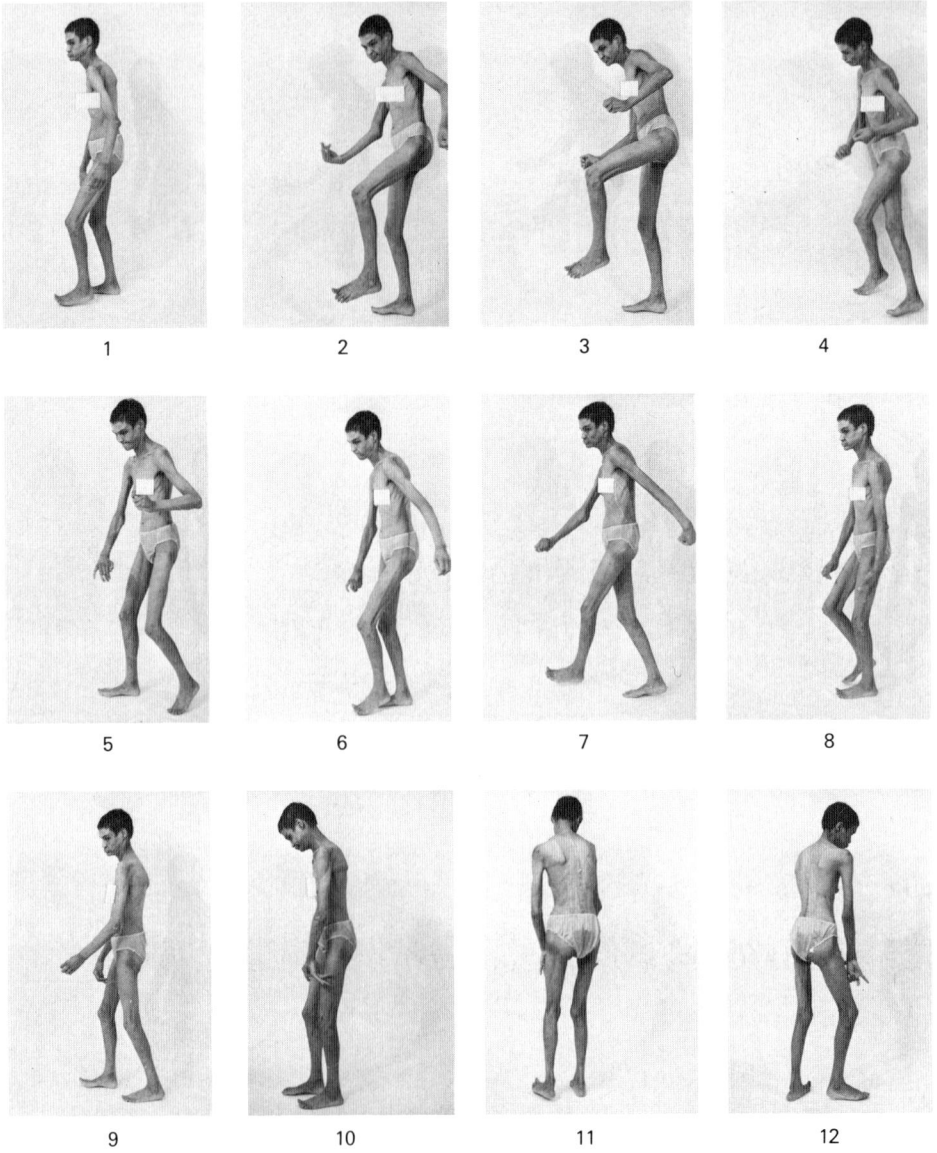

Fig.5.9. Patients with Huntington's chorea have difficulty in walking and may mistakenly be thought to be drunk. (Photographs were taken at 1-s intervals)

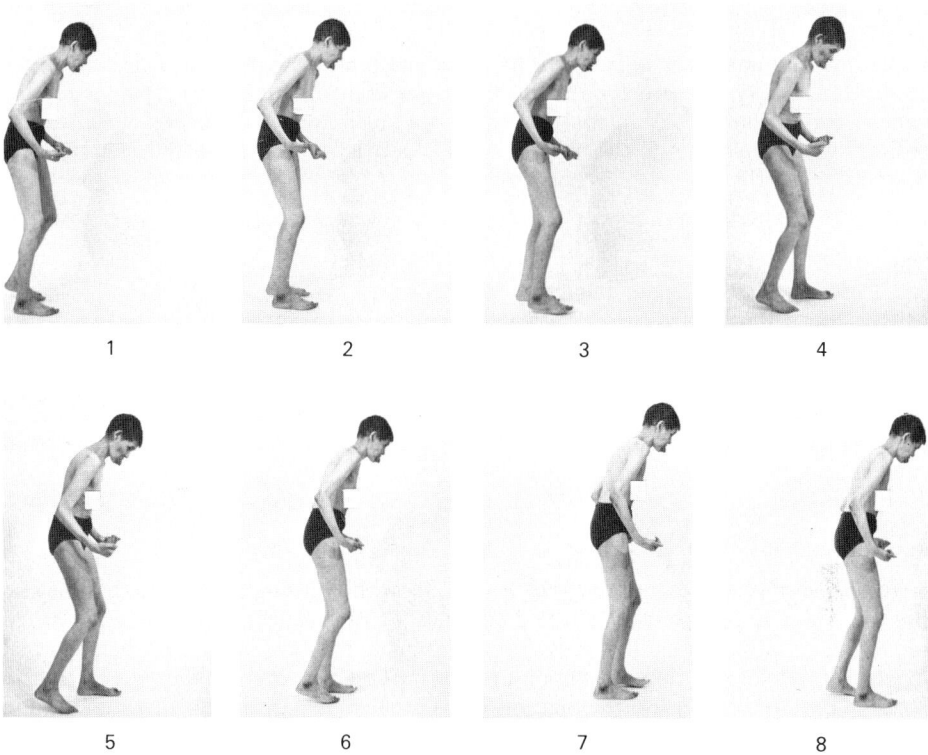

Fig.5.10. Rigid patients have a slow gait with a flexed posture. (Photographs were taken at 2-s intervals)

5.3.6. Oculomotor Dysfunction

Impairment of eye movements is easily discernible in over 50% of all patients with Huntington's chorea. The figure nears 80% if sophisticated methods are used to test for this abnormality.

The oculomotor disorder occurs in both patients with chorea and those with rigidity and is characterised by a selective defect of rapid movements (Starr 1967). Patients requested to look from side to side whilst keeping the head still, show slowness of following movements. Optokinetic nystagmus is also often abnormal with a reduction in velocity of the fast phase. Petit and Milbled (1973) and Avanzini et al. (1979) have studied these defects with the aid of different electro-oculographic recordings. Both groups of workers have documented a significant reduction of saccadic velocity, with vertical saccades being much more impaired than the horizontal. Avanzini and his co-workers have suggested that there is a correlation in Huntington's chorea between the difficulty in performing repeated ocular movements and the impaired execution of gestural sequences, and that this association may point to abnormalities in the frontal cortex or its connections as the pathological basis of this sign.

Changes in the optic nerve are not a feature of the disease, and the pupillary changes (Merskey 1958) and optic atrophy (Stevens and Parsonage 1969) reported previously are probably coincidental findings.

5.3.7. Epilepsy

Grand mal seizures occur in from 1% to 3% of adults with Huntington's chorea (Entres 1940; Hayden 1979; Oepen 1963). This figure is marginally higher than that seen in the general population (0.5%). In contrast, there is a much greater frequency of seizures in patients with early onset of the disorder. This disparity will be discussed more fully in Sect. 5.7.

5.3.8. Incontinence

Approximately 20% of all patients are incontinent of urine and faeces in the terminal phases of the illness and many, in addition, have defective flatus control. However, inability to regulate bowel and bladder function is unusual in persons who have had symptoms for less than 5 years. The underlying pathological basis for impaired sphincter function in this disease in uncertain, but could possibly be attributed to involuntary movements of the controlling muscles, bilateral pyramidal tract dysfunction or frontal cortical atrophy.

5.3.9. Cerebellar Signs

Cerebellar dysfunction is extremely rare in adults with Huntington's chorea but may be present in a small proportion of affected juveniles. Paulson (1979) has alerted clinicians to the difficulty of evaluating this system in Huntington's chorea in view of the abnormal movements which may interfere with normal examining procedure of the cerebellum and result in misinterpretation of the findings. This problem will also be further discussed in Sect. 5.7.

5.3.10. Other Neurological Signs

Muscle wasting of the small muscles of the hand has been commented upon by Panse (1942) and more recently by Stevens (1976), but the pathological basis of this finding is uncertain. There have been no reports of obvious sensory defects in Huntington's chorea. Primitive reflexes such as the grasping or sucking reflex may be elicited as a late consequence of cortical atrophy in some patients.

Purdon Martin (1967) has established that postural reflexes are impaired in approximately 25% of persons with Huntington's chorea. Evidence for this is the loss of tilting reaction and failure to maintain the head in a fixed posture when the patient is blindfolded. There have been no further studies in this regard.

5.4. Mental Disturbance

Whilst chorea is the most obvious and striking clinical feature, it is the mental disturbance in Huntington's chorea which defies medical therapy and is most incapacitating. Some form of mental change is present in all affected persons and although most reports have alluded to these symptoms, there have until recently been few attempts to accurately define and categorise these changes.

5.4.1. Dementia

A major difficulty has arisen with regard to etymology. Different workers have made liberal use of the term 'dementia' without giving a concise explanation of its meaning. The first person to employ this word in connection with Huntington's chorea was Waters (1842), who stated that "it gradually induces a state of more or less perfect dementia". It is likely that he was referring to the obvious and progressive cognitive impairment in the disease. Huntington himself in 1872 mentioned that individuals with the disease have a "tendency to insanity which... sometimes leads to suicide". It is apparent that he was describing the affective disturbance of the disorder. These early descriptions thus embraced two components of the mental changes in Huntington's chorea, namely the cognitive and affective elements. Other common features of the mental disturbance include a personality disorder and a schizophreniform psychosis.

The term 'dementia' will be used here to denote the cognitive impairment, usually reflected in the decline of intellect, the disturbance of memory and reduced capacity for conceptual thought. In the initial phases of the disease patients may show a pattern of focal cognitive impairment (Butters et al. 1978). Whilst intellectual function may be normal as measured by intelligence quotients, disturbance in memory generally occurs early. Memory failure is characterised by a limitation in acquiring new information and severe problems in retrieval of details from short- and long-term memory (Caine et al. 1978). The disturbance of memory is followed by a progressive and severe deterioration of verbal, spatial, arithmetic and perceptual functions.

At the present time it is unclear whether the mental defects described are specific to Huntington's chorea or common to other presenile dementias. Aminoff et al. (1975) have suggested that the mental changes in Huntington's chorea are not qualitatively different from those seen in ageing. Differentiating features between the dementia of Huntington's chorea and Alzheimer's disease are the relative absence of aphasia and agnosia in the former, compared with their invariable presence in the latter. In contrast to the dementia in Huntington's chorea, that in Korsakoff's syndrome centres mainly on short-term memory loss with insignificant intellectual impairment (Butters et al. 1976), whilst that in Parkinson's disease differs in its lower frequency and much later age of onset (Lieberman et al. 1979).

The precise nature of the cognitive changes in Huntington's chorea awaits definition. Furthermore, the pathological basis for these disturbances is unclear. Weingartner et al. (1979) have suggested that the hippocampus may be implicated in view of its known association with memory function (Jaffard et al. 1977), whilst Fedio et al. (1979) have drawn attention to the possible involvement of the corpus callosum and the frontostriatal system in the mental dysfunction of this disorder.

5.4.2. Affective Disturbance

The most common affective disturbance in Huntington's chorea is depression. It is difficult to determine its frequency owing to its multiple modes of presentation. However, suicide and attempted-suicide rates are one crude measure. The phenomenon of the high frequency of suicide in Huntington's chorea will be further discussed in Sect. 8.1.2. McHugh and Folstein (1975) have reported a high frequency of manic-depressive disorder in Huntington's chorea, which has not been confirmed in numerous other surveys (Bolt 1970; Heathfield 1967).

Table 5.3. Frequency of schizophreniform psychosis

Author(s)	No.	Affected population	Frequency (%)
Minski and Guttmann (1938)	3	50	6
Brothers (1964)	16	312	5.1
Heathfield (1967)	9	96	10
Bolt (1970)	58	718	8
Oliver (1970)	12	100	12
Stevens (1976)	12	106	11

5.4.3. Change of Personality

Personality change is an invariable accompaniment of Huntington's chorea and ushers in the disease in almost every instance. In some patients their premorbid personality is exaggerated, with the quiet individual becoming more introverted whilst the more extroverted may become aggressive and violent. In other instances the natural personality trends are completely reversed.

It is uncertain whether the personality disturbance is solely reactive in nature or whether it is also partly an inevitable result of the specific brain damage. It is highly likely that both these factors contribute to this disturbance.

5.4.4. Schizophreniform Psychosis

The frequency of a schizophreniform psychosis preceding or accompanying the abnormal movements in Huntington's chorea varies from 5 to 12% in different reports (Table 5.3). The delusions in these patients are usually paranoid in nature and stories relating to their respective spouses' infidelity or attempts by their partners to kill them are not uncommon. Auditory and visual hallucinations are less frequent.

There is no obvious relationship between this presentation and other clinical features such as chorea or dementia. The psychotic episodes may occur before, concurrently with and after the appearance of these other signs. From different reports it seems, however, that the schizophreniform psychosis is more commonly seen in younger patients with Huntington's chorea.

Garron (1973) has attempted to determine whether this form of psychosis is intrinsically related to Huntington's chorea or whether its simultaneous appearance is a chance event. Schizophrenia occurs in approximately 1% of the population, and the much higher frequency of schizophreniform psychosis occurring in association with Huntington's chorea suggests that this is indeed an associated feature of the disease.

5.4.5. Other Psychiatric Symptoms

Psychosomatic complaints, including headaches and chest and abdominal pain, are occasional accompaniments during the early phases of Huntington's chorea. These symptoms are evidence of latent anxiety which is common in the first stages of the illness.

The understanding of the mental disturbance in Huntington's chorea is still incomplete. Many of the reports have been based on small sample numbers where variables such as class, environment, age and sex have been unspecified. The lack of control groups and the failure to define the clinical characteristics of the patients and the

Table 5.4. Functional designations of patients with Huntington's chorea

Stage	Engagement in occupation	Score	Capacity to handle financial affairs	Score	Capacity to manage domestic responsibility	Score	Capacity to perform activities of daily living	Score	Care can be provided at	Score
Stage 1	Usual level	3	Full	3	Full	2	Full	3	Home	2
Stage 2	Lower level	2	Requires slight help	2	Full	2	Full	3	Home	2
Stage 3	Marginal	1	Requires major help	1	Impaired	1	Mildly impaired	2	Home	2
Stage 4	Unable	0	Unable	0	Unable	0	Moderately impaired	1	Home or extended care facility	1
Stage 5	Unable	0	Unable	0	Unable	0	Severely impaired	0	Total care facility only	0

From Shoulson and Fahn (1979).
Range of scores: 0–13.

stage of their disease are further weaknesses. It is important to recognise these potential problems when planning further investigations.

5.5. Staging

The necessity for a staging system in Huntington's chorea has become obvious. This need has been particularly felt by those investigators who are systematically attempting to define the clinical features and course of Huntington's chorea and by those who are rationally evaluating the efficacy of different forms of management.

There are at least three possible methods for the staging of any disease. The clinical staging system is based on a scoring system for relevant signs. In Huntington's chorea, counting the number of involuntary movements within a given time period could result in a chorea score. A dementia score could be determined by asking a patient questions scaled in difficulty, whilst a similar system could be devised for rigidity. The total score would be an indication of the severity of the disease.

Shoulson and Fahn (1979) have proposed another system, which is based on the patient's functional ability (Table 5.4). This method is particularly useful as it enables clinicians to subdivide their patients according to their social adjustment and productivity, and thereby assess their need for supportive community resources. It also facilitates an examination of the effects of different forms of therapy on the functional status of affected individuals.

The third possible method for staging Huntington's chorea is biochemical, which would allow categorisation of patients according to the severity of and differences in their biochemical disturbance. This type of staging system awaits a more precise understanding of the fundamental biochemical defects of the disorder.

A clinical scoring system in Huntington's chorea has not been devised. The proposed functional staging scheme is the only system available and may serve as a most useful adjunct in the management of patients with Huntington's chorea.

Table 5.5. Landmarks in the description of the different variants of Huntington's chorea

Lyon (1863)	Described two affected children with Huntington's chorea
Westphal (1883)	Described a young patient with rigidity and termed the disorder 'pseudosclerosis'
Gray (1892)	Reported a patient with congenital chorea
Davenport and Muncey (1916)	Introduced the biotype concept. Reported patients with Huntington's chorea sine chorea
Kehrer (1928) } Curran (1930) }	Further reports of Huntington's chorea without choreiform movements
Patzig (1935)	Coined the term 'status subchoreaticus' to delineate patients with minimal chorea and late onset
McKenzie van der Noordaa (1960)	Published an article entitled 'The Westphal Variant of Huntington's chorea'
Bittenbender and Quadfasel (1962)	Delineated four types: the bradykinetic, hyperkinetic and hypokinetic patients and those without movements
Bruyn (1968)	Used the phrase 'juvenile type' to denote onset prior to adulthood, and reviewed prior variants

5.6. Variants

There is at present confusion concerning the delineation of the variants of Huntington's chorea. Davenport and Muncey (1916) first introduced this concept when they used the term 'biotypes' to describe affected families who could be separated by virtue of their differences in age at onset and degree of motor and mental disturbances. Since that time there have been numerous attempts to classify Huntington's chorea and the most important landmarks in this regard are shown in Table 5.5.

The purpose of delineating Huntington's chorea would be to differentiate between certain subtypes and determine their clinical features, natural history and genetic characteristics in the hope that this would enhance our understanding of the disease. However, to achieve this end, the variants ideally must be distinctive and easily differentiated from other forms of the disorder. In this light it is necessary to re-examine the currently recognised variants.

5.6.1. The Westphal Variant

The Westphal variant refers to patients with a degree of rigidity. Bruyn (1968) and Stevens (1976) have pointed out the historical origins of this eponym, which commemorates Westphal who in 1883 described an 18-year-old rigid patient. Whilst Westphal noted that other members of the patient's family, including his father, aunt and uncle, were suffering from chorea, he failed to realise that this was another manifestation of the same disorder. Mistakenly he thought the illness in his patient to be different and termed it 'pseudosclerosis'. The designation of the eponym 'Westphal' to rigid patients is therefore technically incorrect, as he himself did not recognise that rigidity was a feature of Huntington's chorea. This term does not describe a specific, distinctive, non-overlapping form of the disease and for the sake of nosologial clarity could be discarded in favour of an accurate description of the clinical features of the patient concerned.

5.6.2. *Juvenile Huntington's Chorea*

The other well-recognised variant is the 'juvenile' type which, by convention, has been used to denote patients with onset of the disease before the age of 20. This group can be further subdivided into those with signs and symptoms of Huntington's chorea before the age of 10, i.e. childhood onset, and those who present in the second decade, namely with adolescent onset. This classification is extremely arbitrary but is of great potential value, as patients with onset before the age of 20 seem as a group to have different clinical and epidemiological features and an anomalous mode of inheritance of the abnormal gene. In view of this, a description of the clinical features of juvenile Huntington's chorea is warranted. Analysis of the characteristics of this group of patients may clarify the reasons for their differences from adult patients.

5.6.3. *Other Variants*

Other variants which have been described include the 'status sub-choreaticus', which pertains to patients with onset very late in life or in a considerably mitigated fashion. The designation of this one expression for two clinical presentations, which may be distinctly different, is also misleading. For example, some patients may have late onset with severe clinical signs. Other attempts to define variants according to the presenting feature or the absence of chorea (chorea Huntington sine chorea) have been made. None of these classifications has gained acceptance as they are not distinct forms of the disease and have overlapping features with other variants.

5.7. The Clinical Features of Juvenile Huntington's Chorea

Lyon (1863) was the first to record a patient with juvenile Huntington's chorea when he described two children (sibs) in whom well-defined chorea had existed for years. Huntington in his classic paper of 1872 failed to recognise this precocious form of the disease and stated emphatically that a peculiarity of the disorder is that "it manifests itself

Table 5.6. Comparison of reports of juvenile Huntington's chorea

| Author(s) | No. of patients | Predominant motor abnormality | | | Dementia | Epilepsy | Cerebellar signs |
		Chorea	Rigidity	Combined			
Jervis (1963): own series	4	1	3		4	4	1
Jervis (1963): from literature	17	4	8	5	13	7	?
Markham (1969)	9	3	6		7	5	4
Byers et al. (1973)	4		4		4	4	2
Hayden (1979)	17	4	10	2	16	6	3
Total	51	12 (23.5%)	31 (60.7%)	7 (13.7%)	44 (86.2%)	26 (50.9%)	10 (19.6%)

only in adult life". Since that time over 175 separate papers concerning juvenile Huntington's chorea have been published, most of which have been single case reports.

There is a striking clinical disparity between the features of Huntington's chorea in children and those in adults, with particular regard to the principal motor abnormalities, the rate of progression, the occurrence of epilepsy and cerebellar signs. The results obtained in some of the largest reported series are summarised in Table 5.6.

Chorea is generally an early sign, mild in degree and usually superseded by rigidity, which is the predominant motor abnormality (Fig. 5.11a,b,c). The rigidity is initially extrapyramidal in nature but in most instances is later accompanied by signs of pyramidal tract dysfunction. Gegenhalten, the increasing resistance to movement seen in frontal lobe disorders, is quite frequently seen in established cases of Huntington's chorea.

Intellectual deterioration with failure at school and difficulty in concentration is found in almost all patients and is often the presenting feature. The clinical features of 17 such patients from South Africa are shown in Table 5.7. Other mental aberrations are not uncommon and a psychosis resembling paranoid schizophrenia was the presenting feature in 3 of the 17 listed patients.

Seizures occur in approximately 30%–50% of affected juveniles and are usually a late phenomenon. These are typical grand mal type convulsions, although petit mal seizures have also been reported (Green et al. 1973). Epilepsy is often difficult to control with routine anti-epileptic medication in these patients.

The difficulty in eliciting cerebellar signs has been mentioned previously. Nevertheless there have been repeated reports of such findings, which have included dysmetria, dysdiadochokinesia and intention tremor. Nystagmus is a less frequent feature. It would seem that cerebellar signs may indeed occur in a small proportion of juvenile patients but even in this group they are probably an infrequent occurrence.

Various theories have been invoked to account for the differences in presentation between children and adults with Huntington's chorea. Attempts to correlate the severity of destruction of the caudate nucleus, putamen and globus pallidus with chorea or rigidity have not been conclusive. The morphological abnormalities as a rule conform to the findings seen in affected adults (Goebel et al. 1978). Differences are the more common finding of cerebellar, inferior olivary and globus pallidus lesions in juvenile patients (Jervis 1963; Markham and Knox 1965). The former changes would account for the signs of cerebellar disease, whilst the increased frequency of rigidity could be attributed to the gliosis of the globus pallidus (Byers et al. 1973). However, no pathological differences which might explain the disparity in seizure frequency between adult and juvenile patients have been demonstrated. In summary, the morphological differences are not consistently present and do not satisfactorily explain the varying clinical features.

The increased frequency of epilepsy in affected juveniles may reflect the greater likelihood of seizures in children generally when exposed to a variety of stresses. This is probably not the result of any specific difference in the precocious form of the disease, but rather the pattern of reaction to the degenerative process in the developing brain, as opposed to the developed brain of affected adults. The fact that adults with the rigid form of the disease do not have an increased frequency of epilepsy suggests that the occurrence of seizures is unrelated to the clinical presentation of rigidity.

The shorter duration of juvenile Huntington's chorea may also reflect the process of degeneration in a brain still undergoing growth and development. An immature central nervous system may be less able to withstand the destructive process of Huntington's chorea than is the adult brain, with consequent shorter duration of the disease.

Juvenile Huntington's chorea is not a common disease but it may be more common than is generally supposed. The lack of awareness of this form of the disorder has resulted

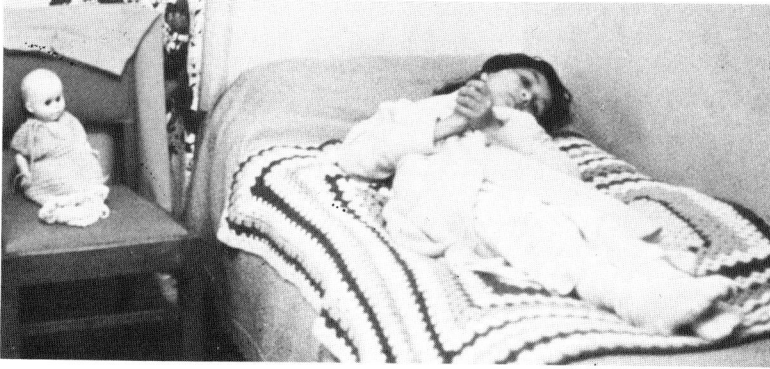

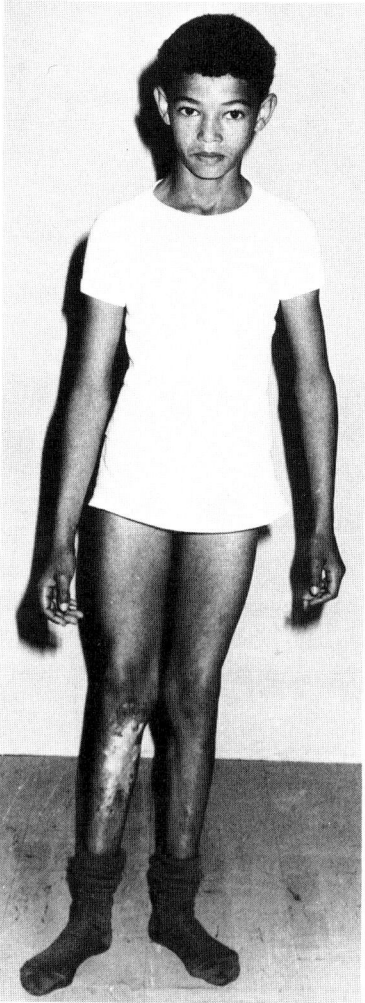

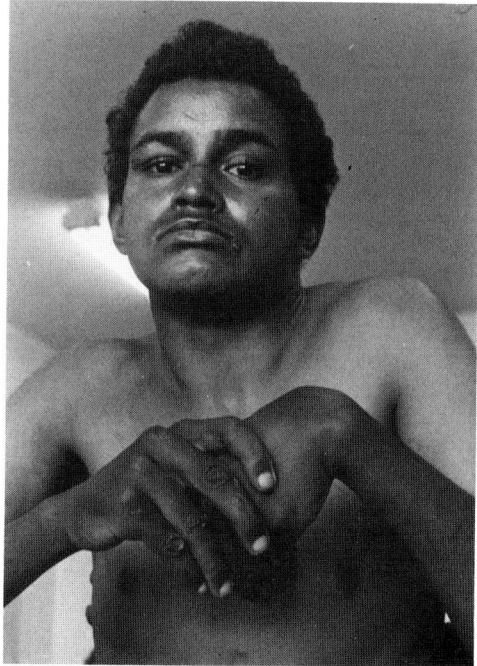

Fig. 5.11. a A girl aged 15, two months before her death, with flexed arms and hyperextended legs and feet

b This 14-year-old boy had onset at 8 years with intellectual deterioration and mild chorea. Rigidity is now the major sign. The scar on his right shin is a result of his fall into a fire at 10 years of age

c A schizophreniform psychosis heralded the onset of Huntington's chorea at 15 in this 20-year-old male. His predominant motor abnormality is also rigidity

Table 5.7. Clinical features of 17 patients with juvenile Huntington's chorea in South Africa

Initials	Sex	Sex of affected parent	Age of onset/ death	Presenting symptoms	Chorea	Rigidity	Dementia	Cerebellar signs	Epilepsy	Initial diagnosis
Mixed ancestry										
V.v.W.	M	M	14	Mental impairment	**	***	**	*	**	Dementia
M.v.W.	F	M	12/19	Clumsiness, mental impairment	***	*	***	—	*	Huntington's chorea
B.v.W.	M	M	16	Mental impairment	—	***	*	*	—	Huntington's chorea
K.v.W.	F	M	15	Mental impairment	*	*	*	—	—	Huntington's chorea
E.M.	F	M	8/16	Clumsiness, mental fall-off	*	***	**	—	**	Hyperactivity
E.M.	F	M	12	Psychosis	—	—	**	—	**	Schizophrenia
I.M.	F	M	5/15	Dysarthria	**	*	**	—	—	Infantile autism
C.D.	M	M	16	Psychosis	*	**	*	—	—	Schizophrenia
L.D.	M	F	19	Psychosis	*	—	*	—	—	Schizophrenia
L.J.	M	F	8	Social withdrawal	—	**	***	—	—	Minimal brain damage
J.S.	F	M	17	Irritability	*	**	*	—	—	Wilson's disease
White										
D.D.	F	M	9/15	Mental fall-off	*	***	**	—	**	Unknown
A.H.	F	M	5/15	Mental fall-off	*	***	***	*	**	Mental retardation
S.S.	F	F	10	?	**	—	*	—	—	Unknown
S.S.	F	F	15	?	*	*	—	—	—	Unknown
J.V.	M	M	13/19	?	**	*	*	—	—	Unknown
J.d.T.	M	M	5	?	*	**	—	—	—	Unknown

*, mild; **, moderate; ***, severe.

in a very high frequency of initial misdiagnosis, which has caused further problems for affected families. An index of suspicion is fully justified in the case of any child who presents with unexplained psychiatric or neurological symptomatology, if a positive family history of the disorder has been obtained.

5.8. Diagnostic Techniques

There is no available diagnostic technique which yields results pathognomonic and absolutely specific to Huntington's chorea. The role of special investigations in this disorder is to confirm the clinical diagnosis. In a few situations where diagnostic uncertainty exists, these techniques may be of special use.

The most definitive method is by post-mortem examination of the affected brain, and the characteristic pathological features will be discussed in Chap. 6. Even with this method some diagnostic difficulty may ensue. The model procedure for the diagnosis of Huntington's chorea should be distinctive, with no overlap in results between this disorder and other diseases or normal persons, and would ideally show no false positives or false negatives. At present no such technique exists. The diagnostic procedures that have been used will now be reviewed.

5.8.1. Electroencephalographic Studies (EEG)

There have been numerous accounts of the EEG findings in Huntington's chorea. The most common reported abnormality has been a low-voltage, poorly formed EEG. The most comprehensive study was that conducted by Scott et al. (1972) on 95 patients. One-third (31) of affected individuals in that investigation had an EEC with no alpha rhythm over 10 mV in amplitude. The remaining 64 patients either had a normal EEG or showed non-specific changes. Other findings have confirmed the high frequency of false negatives, which may be of the order of 70% with this technique.

In addition, whilst the low-voltage EEG is not common in other disorders it is not specific for Huntington's chorea and thus is not useful as an aid to the diagnosis.

5.8.2. Pneumoencephalographic Studies (PEG)

The PEG may still be of great value in substantiating the diagnosis of Huntington's chorea. However, as it is an invasive procedure, with a significant morbidity, it should not be used routinely and can only be justified if the results offer essential diagnostic information.

The cardinal pathological feature of Huntington's chorea is caudate nucleus atrophy. This results in dilatation of the lateral ventricles which is easily discernible on PEG (Fig. 5.12a,b,c). Another important feature of the PEG of affected persons is the widening of the cerebral sulci, which is the consequence of cortical atrophy. These PEG findings are present to different degrees in most but not all patients (about 90%). Whilst all patients with advanced disease have some abnormalities, these changes do not as a rule correlate with the severity or the duration of the illness.

Different indices have been devised in an attempt to establish criteria for the diagnosis. Gath et al. (1975) have devised an index based on the frontal horn width divided by the septate–caudate distance, which has a low average value in Huntington's chorea. However, the suggested diagnostic value of 1.40 for this ratio (Gath and Vinje 1968) shows considerable overlap with Parkinson's disease and generalised cerebral atrophy

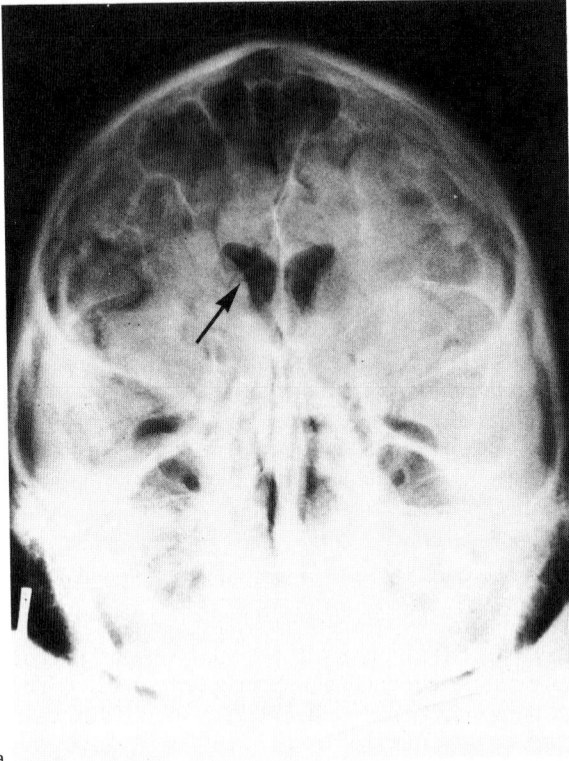

a

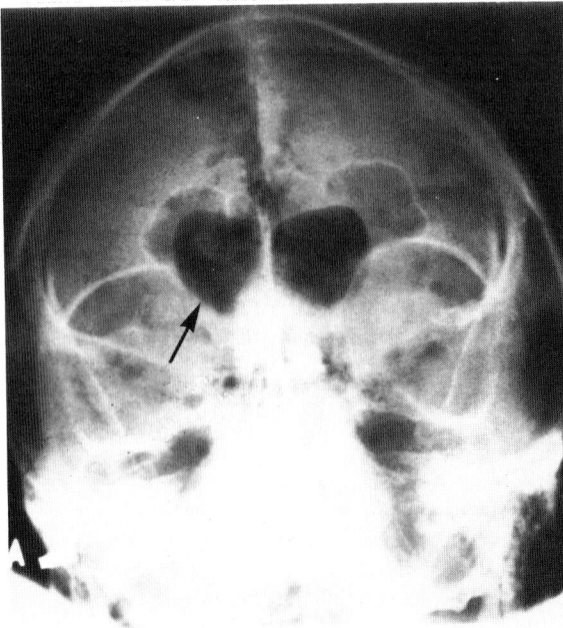

b

Fig. 5.12. a A normal pneumoencephalogram showing the
characteristic indentation due to the
caudate nucleus
  b A PEG of a 53-year-old man with
advanced Huntington's chorea,
showing ventricular dilation
consequent to caudate nucleus atrophy

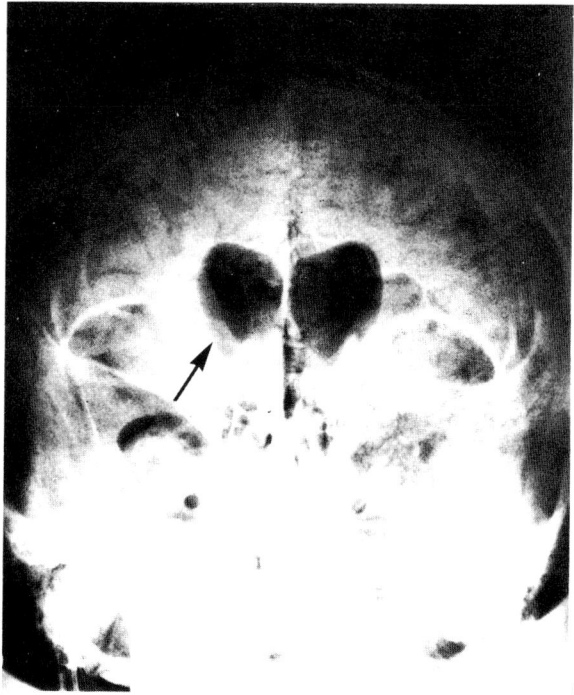

Fig.5.12. c A PEG of a 15-year-old
girl with marked rigidity, showing
similar features

c

(Fahn et al. 1973). Pick and Wilson's disease may also give identical PEG patterns
(Blinderman et al. 1964).

For these reasons the PEG is not an ideal test and has been superseded by the CAT
scan.

5.8.3. Computerised Axial Tomography (CAT)

The advent of the CAT scan has offered a non-invasive, relatively harmless method for
assessing ventricular dilatation. Terrence et al. (1977) were the first to document results
in this regard. Objective measurements of CAT scans of 12 patients with Huntington's
chorea were found to be statistically different from results of scans of normal persons and
those with cerebral atrophy and Parkinson's disease.

Two CAT scan measurements are particularly useful in assessing caudate nucleus size:

1) the FH line, which is defined as the largest frontal horn span; and

2) the CC line, or bicaudate diameter, which is defined as the shortest transverse
distance between the medial borders of the caudate nuclei (Fig. 5.13).

Measurements are made in millimetres. The CC line increases in Huntington's chorea
due to degeneration of the corpus striatum (Fig. 5.14). As a result the FH/CC ratio
decreases. Neophytides et al. (1979) used this ratio in their study of 42 patients and found
that the mean of 1.56 ± 0.1 was lower than in the normal population. However, two

affected patients with intellectual deterioration but no chorea had ratios within normal limits and close examination of individual results showed a broad scatter with overlap between Huntington's chorea ratios and those of normal persons. One possible factor also to be considered in this connection is that there is a general increase in ventricular size with age.

Barr et al. (1978) have proposed another index, namely the bicaudate index. The distance between the outer tables of the skull at the level of the bicaudate diameter is designated OT_{CC}. OT_{FH} is the distance between the outer tables of the skull at the level of the FH line as defined previously. The bicaudate index is the ratio of the bicaudate diameter (CC) to the distance between the outer tables of the skull at that level (Fig. 5.13).

This index significantly discriminates Huntington's chorea patients from normal persons and others with cortical atrophy from an unrelated cause. In their small series of 7 patients with Huntington's chorea, 20 persons with cerebral atrophy and 20 controls, Barr et al. found that the bicaudate index allowed better differentiation than the FH/CC ratio. Whilst there was a lack of strict correlation between radiographic findings and clinical symptoms, a more severe dilatation often occurred in the more advanced cases.

Computerised tomography is an important advance in the available diagnostic techniques for Huntington's chorea. Unfortunately, false negatives in patients with little or no chorea have already been reported and it is just in those instances that a diagnostic tool is needed. Nevertheless, refinement of techniques and equipment may obviate these problems and enhance the usefulness of the CAT scan as a diagnostic aid in Huntington's chorea.

5.8.4. Cerebral Angiography

Fahn et al. (1973) have reported a consistent decrease in the calibre and number of blood vessels supplying the neostriatum in patients with Huntington's chorea compared with the normal distribution of these vessels in diffuse cerebral and Parkinson's disease. This technique with its hazards offers no advantage over the diagnostic methods previously described.

5.9. Problems of Diagnosis

The protean clinical symptomatology of patients with Huntington's chorea has resulted in frequent misdiagnosis. This occurs particularly in persons with unusual clinical presentations and in the setting of a negative family history. Problems of diagnosis are compounded by the relative rarity of the disorder, which means that the majority of doctors have had little experience of the disease.

5.9.1. Misdiagnosis

Misdiagnosis of Huntington's chorea can occur in two ways. Firstly, patients with this disorder may be misdiagnosed as suffering from some other disease. Secondly, persons with some other condition may be wrongly diagnosed as being affected with Huntington's chorea.

Although less common, the latter situation is particularly serious in view of the important clinical and far-reaching genetic implications of Huntington's chorea. In this context the importance of the post-mortem examination in confirming the diagnosis will be discussed in Chap. 6.

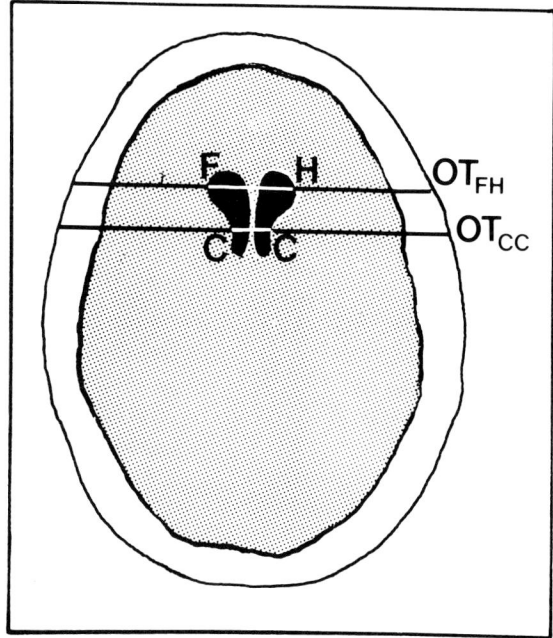

Fig.5.13. Diagram of the CAT scan measurements useful for assessing caudate nucleus size

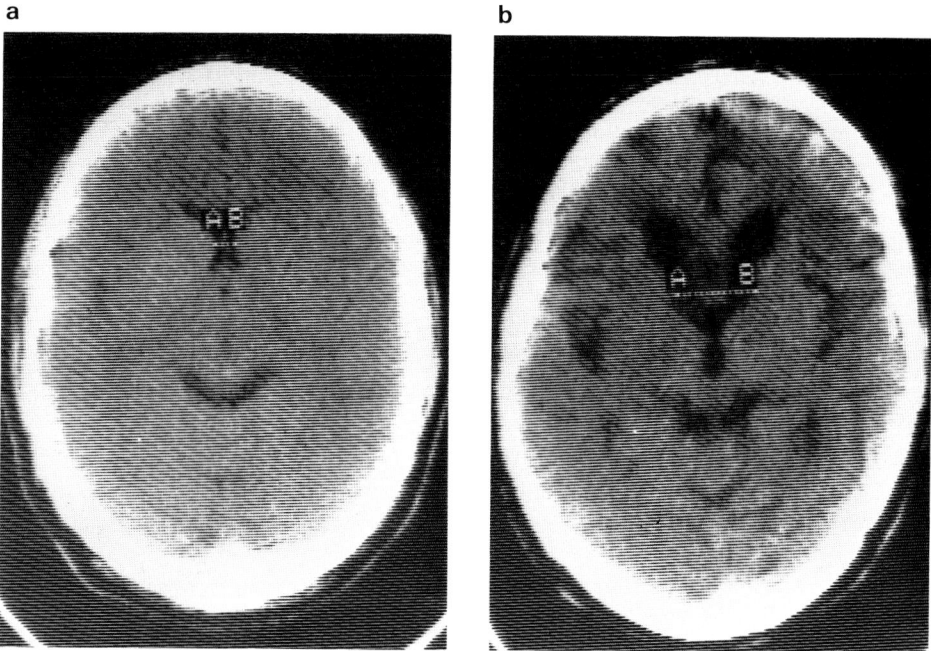

Fig.5.14. CAT scan of a person with Huntington's chorea (**b**) compared with that of an age-matched normal control (**a**). The frontal horn diameter and bicaudate diameter are obviously increased. Cortical atrophy is evident. (By courtesy of David Stevens)

Table 5.8. Differentiating characteristics between Huntington's chorea and Alzheimer's disease

Characteristics	Huntington's chorea	Alzheimer's disease[a]
Age of onset	± 35 years	± 55 years
Genetics	Autosomal dominant (AD)	Multifactorial AD occasionally
Duration	± 14 years	± 8 years
Extrapyramidal features	Chorea in ± 90% (adults)	Parkinsonism in ± 70%
Aphasia	Rare	± 90%
Astereognosia	Rare	± 70%
Epilepsy	1% of adult patients	± 14%

[a]From Pearce and Miller (1973).

The condition most commonly mistaken for Huntington's chorea is Alzheimer's disease. Confusion may arise as both disorders may be inherited as autosomal dominant traits and may present with a change in personality. However, there are important differentiating features, which are summarized in Table 5.8. It is noteworthy that there has been one report of the unusual coincidence of Huntington's chorea occurring simultaneously with Alzheimer's disease (McIntosh et al. 1978).

The wrong diagnosis of Huntington's chorea may also be made in patients with spinocerebellar ataxia and tardive dyskinesia. The differentiation between Huntington's chorea and tardive dyskinesia may on occasion be extremely difficult. However, tardive dyskinesia is a unique entity with specific motor abnormalities. The clinical discriminating characteristics are shown in Table 5.9. A history of drug usage should be made in every patient suspected of having Huntington's chorea. Whilst the antipsychotic drugs (phenothiazines, butrophenones) are the major causative agents of tardive dyskinesia, all drugs which alter catecholaminergic status, such as L-dopa, other dopamine agonists, phenytoin and oral contraceptives, may induce chorea.

Table 5.9. Differentiating characteristics between Huntington's chorea and tardive dyskinesia

Characteristics	Huntington's chorea	Tardive dyskinesia[a]
Dyskinesia	May affect all parts of the body	Predominantly a repetitive, bucco-lingual-masticatory syndrome
Dysarthria	Common, severe	Infrequent, mild
Extremities	Involves proximal and distal segments	Usually only distal
Gait	Characteristically abnormal	Usually normal
Tremor	Uncommon	In about 15-50% of patients
Course	Progressive	Static or fluctuant
Dementia	Usual	Unusual

[a]From Crane (1973).

The more common situation, however, is that patients with Huntington's chorea are misdiagnosed as having other neurological or psychiatric conditions, most commonly schizophrenia. The advent of choreiform movements in these patients is often regarded as a schizophrenic mannerism or the side effect of neuroleptic therapy, and thus the wrong diagnosis persists. In some instances it is only some years later, when the patient's deterioration is noted and a positive family history of Huntington's chorea emerges, that this mistake is corrected. Other common misdiagnoses are depression, Parkinson's disease, multiple sclerosis and other organic brain syndromes.

The confusion between Huntington's chorea and other conditions is compounded by the fact that at-risk and affected individuals may suffer from other neurological or psychiatric illnesses. Furthermore, such persons may also develop chorea in association with some other disease. Chorea is a non-specific sign of basal ganglia dysfunction due to many different causes. In addition other diseases not thought to directly affect the basal ganglia, such as thyrotoxicosis, lupus erythematosis, polycythaemia and many others, may also present with this abnormal movement (Godwin-Austen 1979; Greenhouse 1966). Chorea may also be a late complication of a subdural haematoma (Gilmore and Brenner 1979). In practice most of these other diseases have obvious distinguishing characteristics which will make their differentiation from Huntington's chorea relatively easy.

A major theme throughout this chapter has been that a positive family history is most important for the diagnosis of Huntington's chorea. Whilst this is absolutely correct, it must be realised that there are many causes of inherited chorea, only one of which (albeit the commonest) is Huntington's chorea.

5.9.2. The Differential Diagnosis of Inherited Chorea

Hereditary chorea is not synonymous with Huntington's chorea and the differentiation between the multiple causes of inherited chorea is most important in terms of their different modes of inheritance, course, treatment and prognosis. The more common causes of inherited chorea and their clinical characteristics are shown in Table 5.10.

There are five important causes of autosomal dominant inherited chorea. Haerer et al. (1967) were the first to document *benign hereditary chorea* in 15 members of a Black American kindred. Since that time there have been approximately ten such reports confirming that this is a distinct entity, easily distinguished from Huntington's chorea by its onset in infancy, non-progressive course and amelioration with age. Different terms, including familial non-progressive chorea (Bird et al. 1976) and familial essential chorea have been used to describe this disease.

Recently Fisher et al. (1979) have described a similar disorder, but which has disproportionate involvement of the legs and a gradual increase in severity. They have termed the disorder *familial inverted choreoathetosis*. Further reports are needed to confirm its differentiation from benign hereditary chorea prior to its universal acceptance as a distinct entity.

Mount and Reback (1940) were the first to describe *familial paroxysmal choreoathetosis*, which is clearly different from these other disorders by the intermittent nature of the symptoms which are possibly epileptic in origin. Estes et al. (1967) reported a 'new *hereditary acanthocytosis syndrome*' in nine members of a kindred, all of whom had some neurological defect, usually chorea in the presence of acanthocytes in the peripheral blood smear. Both autosomal dominant (Estes et al. 1967; Levine et al. 1968; Kito et al. 1980) and autosomal recessive forms of the disorder have been described (Bird et al. 1978; Critchley and Nicholson 1970). This syndrome is dissimilar to Huntington's chorea in

Table 5.10. Common causes of inherited chorea

	Inheritance[a]	Age at onset	Course	Clinical features	Associated findings
Huntington's chorea	AD	About 35	Progressive	Chorea, dementia	Rigidity
Familial paroxysmal choreoathetosis	AD, AR	Childhood	Non-progressive	Bouts of choreoathetosis	Abnormal EEG; responds well to anti-epileptic therapy; precipitated by coffee, fatigue, tiredness
Familial degeneration of basal ganglia with acanthocytosis	AD, AR	Early adulthood	Slowly progressive	Chorea	Late-onset dementia; hyporeflexia; 5%–15% oro-facial dyskinesia; acanthocytes; epilepsy; muscle wasting
Benign hereditary chorea	AD, AR	Early childhood	Non-progressive	Chorea, gait difficulty	No dementia; tremor; amelioration with age
Familial inverted choreoathetosis	AD	Before 1 year	Slowly progressive	Choreoathetosis	Mainly affects lower limbs
Wilson's disease	AR	Variable, usually childhood	Progressive	Tremor, dysarthria, rigidity	Cirrhosis; Kayser-Fleisher ring in cornea
Familial calcification of the basal ganglia	AR	Variable, usually adulthood	Non-progressive	Parkinsonian features	Seizure; chorea; dementia; ataxia
Hallervorden-Spatz syndrome	AR	Childhood	Progressive	Dementia, choreo-athetosis	Rigidity
Pallidonigral degeneration	AR	Adolescence	Slowly progressive	Parkinsonian features	Pallidal pigmentation; choreoathetosis; thrombo-cytopaenia; hepato-splenomegaly
Pseudohypoparathyroidism	X-linked dominant	Variable	Variable	Parkinsonian features	Skeletal anomalies; chorea; mental retardation; epilepsy; microcephaly
Lesch-Nyhan syndrome	X-linked recessive	Infancy	Progressive	Choreoathetosis	Self-mutilation; mental retardation; hyperuricaemia

[a]AD, autosomal dominant; AR, autosomal recessive.

terms of the presence of acanthocytes, hyporeflexia and relative lack of mental deterioration.

The other more common causes of inherited chorea are also shown in Table 5.10, and these are clearly distinguishable from Huntington's chorea by their different mode of inheritance and their clinical characteristics.

Other conditions which do not usually cause chorea, but may still be confused with Huntington's chorea, include benign hereditary myoclonus, Gilles de la Tourette's disease and Joseph disease. *Benign hereditary myoclonus* is characterised by abrupt onset in childhood of non-progressive sharp involuntary movements, particularly involving the proximal segments of the extremities; it is usually inherited as an autosomal dominant (Daule and Peters 1966). *Gilles de la Tourette's disease* begins in childhood and is initially characterised by shoulder girdle tics, followed by head bobbing, an expiratory bark and coprolalia (Meyerhoff and Snyder 1973). Although genetic factors are implicated, the exact mode of inheritance is not established. *Joseph disease* is an autosomal dominant disorder with onset between the ages of 20 and 50 and with ataxia, ophthalmoplegia and pyramidal signs as the major clinical features. It was first described in patients who emigrated to California from Flores, Azores Islands, in 1844 and today is also seen in the New England states of the USA (Coutinho and Andrade 1978; Rosenberg et al. 1979).

Careful enquiry as to the mode of inheritance, age at onset and course of the disease, with a thorough clinical examination, will enable the clinician to differentiate on most occasions between Huntington's chorea and other causes of inherited chorea.

References

Aminoff MJ, Gross M (1974) Vasoregulatory activity in patients with Huntington's chorea. J Neurol Sci 21: 331-8

Aminoff MJ, Marshall J, Smith EM, Wyke MA (1975) Pattern of intellectual impairment in Huntington's chorea. Psychol Med 5: 169-172

Avanzini G, Girotti F, Caraceni T, Spreafilo R (1979) Oculomotor disorders in Huntington's chorea. J Neurol Neurosurg Psychiatry 42: 581-589

Barr AN, Heinze WJ, Doggen GD, Valvassor GE, Sugar O (1978) Bicaudate index in computerized tomography of Huntington's disease and cerebral atrophy. Neurology 28: 1196-1200

Bird TD, Carlson CB, Hall JG (1976) Familial essential chorea. J Med Genet 13: 357-362

Bird TD, Cederbaum S, Valpey RW, Stahl WL (1978) Familial degeneration of the basal ganglia with acanthocytosis: a clinical, neuropathological and neurochemical study. Ann Neurol 3: 253-258

Bittenbender JB, Quadfasel FA (1962) Rigid and akinetic forms of Huntington's chorea. Arch Neurol 7: 275-288

Blinderman EE, Weidner W, Markham CH (1964) The pneumoencephalogram in Huntington's chorea. Neurol 14: 601-607

Bolt JM (1970) Huntington's chorea in the West of Scotland. Br J Psychiatry 116: 259-270

Brackenridge CJ (1971) The relation of type of initial symptoms and line of transmission to ages at onset and death in Huntington's disease. Clin Genet 2: 287-297

Brothers CR (1964) Huntington's chorea in Victoria and Tasmania. J Neurol Sci 1: 405-420

Bruyn GW (1968) Huntington's chorea: historical, clinical and laboratory synopsis. In: Vinken PJ, Bruyn GW (eds) Handbook of clinical neurology, vol VI. North-Holland, Amsterdam, pp 268-378

Bruyn GW (1973) Clinical variants and differential diagnosis. In: Barbeau A, Chase TN, Paulson GW (eds) Huntington's chorea, 1872-1972. Raven Press, New York, pp 51-56

Butters N, Tarlow S, Cermak LS, Sax D (1976) A comparison of the information processing deficits of patients with Huntington's chorea and Korsakoff's syndrome. Cortex 12: 134-144

Butters N, Sax D, Montgomery K, Tarlow S (1978) Comparison of the neuropsychological deficits associated with early and advanced Huntington's disease. Arch Neurol 35: 585-589

Byers RK, Gilles FH, Fund C (1973) Huntington's disease in children. Neurology 23: 561-569

Caine ED, Hunt RD, Weingartner H, Ebert MH (1978) Huntington's dementia: clinical and neuro-
    psychological features. Arch Gen Psychiatry 35: 377-384
Coutinho P, Andrade C (1978) Autosomal dominant system degeneration in Portuguese families of the Azores
    Islands. Neurology 28: 703-709
Crane GE (1973) Tardive dyskinesia and Huntington's chorea. Drug-induced and hereditary dyskinesias. In:
    Barbeau A, Chase TN, Paulson GW (eds) Huntington's chorea, 1872-1972. Raven Press, New York, pp
    115-122
Critchley EM, Nicholson JM (1970) Acanthyocytosis, normoproteinemia and multiple tics. Postgrad Med J
    46: 698-701
Curran D (1930) Huntington's chorea without choreiform movements. J Neurol Psychopathol 10: 305-310
Daule JR, Peters HA (1966) Hereditary essential myoclonus. Arch Neurol 15: 587-590
Davenport CB, Muncey EB (1916) Huntington's chorea in relation to heredity and eugenics. (Bulletin of
    Eugenics Record Office 17). Carnegie Institute, Washington
Denny-Brown D (1962) The basal ganglia. Oxford University Press, London
Dynan NJ (1913-1914) The physical and mental states in chronic chorea; summary of 19 cases of chronic
    progressive chorea, with post-mortem findings in 8 cases. Am J Insan 70: 589-636
Edmonds C (1966) Huntington's chorea, dysphagia and death. Med J Aust 2: 273-274
Entres JL (1940) Der Erbvietstanz Huntingtonsche Chorea: erbbiologischer Teil. In: Gutt A Handbuch der
    Erbkrankheiten, vol III. Thieme, Leipzig, pp 243-262
Estes JW, Morley T, Levine IM, Emerson CP (1967) A new hereditary acanthocytosis syndrome. Am J Med
    42: 868-881
Fahn S, Mishkin M, Hoffman RR (1973) Pharmacologic and radiologic investigations in Huntington's chorea.
    In: Barbeau A, Chase TN, Paulson GW (eds) Huntington's chorea, 1872-1972. Raven Press, New York, pp
    581-598
Fedio P, Cox CS, Neophytides A, Canal-Frederick G, Chase TN (1979) Neuropsychological profile of
    Huntington's disease: patients and those at risk. In: Chase TN, Wexler NS, Barbeau A (eds) Advances in
    neurology, vol 23. Raven Press, New York, pp 239-256
Fisher M, Sargent J, Drachman D (1979) Familial inverted choreathetosis. Neurology 29: 1627-1631
Garron DC (1973) Huntington's chorea and schizophrenia. In: Barbeau A, Chase TN, Paulson GW (eds)
    Huntington's chorea, 1872-1972. Raven Press, New York. pp 273-285
Gath I, Vinje B (1968) Pneumoencephalographic findings in Huntington's chorea. Neurology (Minneap) 18:
    991-996
Gath I, Jørgensen A, Sjaastad O, Berstad J (1975) Pneumoencephalographic findings in Parkinsonism. Arch
    Neurol 32: 769-773
Gilmore PC, Brenner RP (1979) Chorea: a late complication of a subdural haematoma. Neurology (Minneap)
    29: 1044-1045
Godwin-Austen RD (1979) The treatment of the choreas and athetotic dystonias. J R Coll Physicians Lond 13:
    35-38
Goebel HH, Herpertz R, Scholz W, Igbad K, Tellez-Nagel I (1978) Juvenile Huntington's chorea: clinical,
    ultrastructural and biochemical studies. Neurology (Minneap) 28: 23-31
Goodman RM, Hall CL, Terango L, Perrine GA, Roberts PL (1966) Huntington's chorea: a multidisciplinary
    study of affected parents and first generation offspring. Arch Neurol 15: 345-355
Gray LC (1892) A case of Huntington's chorea: also one of congenital Huntington's chorea. J Nerv Ment Dis
    19: 725-727
Green JB, Dickinson ES, Gunderman JR (1973) Epilepsy in Huntington's chorea: clinical and neuro-
    physiological studies. In: Barbeau A, Chase TN, Paulson GW (eds) Huntington's chorea, 1872-1972.
    Raven Press, New York, pp 105-113
Greenhouse AH (1966) On chorea, lupus erythematosis and cerebral arteritis. Arch Intern Med 117: 389-393
Haerer AF, Currier RD, Jackson JF (1967) Hereditary non-progressive chorea of early onset. N Engl J Med
    276: 1220-1224
Hamilton AS (1908) Report on twenty-seven cases of chronic progressive chorea. Am J Insan 64: 403-475
Hayden MR (1979) Huntington's chorea in South Africa. PhD thesis, University of Cape Town
Hayden MR, Vinik AL (1979) Disturbances in hypothalamic-pituitary hormonal dopaminergic regulation in
    Huntington's disease. In Chase TN, Wexler N, Barbeau A (eds) Advances in neurology, vol. 23. Raven
    Press, New York, pp 305-318
Heathfield KWG (1967) Huntington's chorea: investigation into prevalence of this disease in the area covered
    by the North East Metropolitan Hospital Board. Brain 90: 203-232
Huntington G (1872) On chorea. Med Surg Rep 26: 317-321
Jaffard R, Ebel A, Destrade C, Durkin T, Mandel P, Cardo B (1977) Effects of hippocampal electrical
    stimulation on long-term memory and on cholinergic mechanisms in three inbred strains of mice. Brain Res
    133: 277-289

Jervis GA (1963) Huntington's chorea in childhood. Arch Neurol 9: 244-257

Kehrer FA (1928) Erblichkeit und Nervenleiden. I. Ursache und Erblichkeit von Chorea, Myoklonie und Athetose. Monographien aus dem Gesamtgebiete der Neurologie und Psychiatrie, Heft 50. Springer, Berlin

Kinnier Wilson SA (1954) In: Bruce AN (ed) Neurology, 2nd ed. Butterworth, London, pp 712-714

Kito S, Itoga E, Hiroshige Y, Matsumoto N, Nliwa S (1980) A pedigree of amyotrophic chorea with acanthocytosis. Arch Neurol 37: 514-517

Levine I, Estes JW, Looney JM (1968) Hereditary neurological disease with acanthocytosis. Arch Neurol 19: 403-409

Lieberman A, Dziatolowski M, Neophytides A, Kupersmith M, Aleksic S, Serby M, Korein J, Goldstein M (1979) Dementias of Huntington's and Parkinson's disease. In: Chase TN, Wexler NS, Barbeau A (eds) Advances in neurology, vol 23. Raven Press, New York, pp 273-280

Lyon JW (1863) Chronic hereditary chorea. Am Med Times 7: 289-290

MacKenzie Van Der Noordaa MC (1960) De door Westphal beschreven variant van de ziekte van Huntington. Ned Tijdschr Geneeskd 104: 1625-1627

McHugh PR, Folstein MF (1975) Psychiatric syndromes of Huntington's chorea. In: Benson F, Blumer D (eds) Psychiatric aspects of neurological disease. Grune & Stratton, New York, pp 257-285

McIntosh GC, Jameson HD, Markesberry WR (1978) Huntington's disease associated with Alzheimer disease. Ann Neurol 3: 545-548

Markham CH, Knox JW (1965) Observations on Huntington's chorea in childhood. J Pediatr 67: 46-57

Markham CH (1967) Huntington's chorea in childhood. In: Barbeau A, Brunette JR (eds) Progress in Neurogenetics. Excerpta Medica, Amsterdam, pp 651-659

Mattsson B (1974) Clinical, genetic and pharmacological studies in Huntington's chorea. UMEA University Medical Dissertations 7. UMEA, Sweden, pp 21-51

Merskey H (1958) General paresis complicating Huntington's chorea. J Ment Sci 104: 1203-1204

Meyerhoff JL, Snyder SH (1973) Catecholamines in Gilles de la Tourette's disease: a clinical study with amphetamine isomers. In: Barbeau A, Chase TN, Paulson GW (eds) Huntington's chorea, 1872-1972. Raven Press, New York, pp 123-134

Minski L, Guttmann E (1938) Huntington's chorea: a study of 34 families. J Ment Sci 84: 21-96

Mount LA, Reback S (1940) Familial paroxysmal choreoathetosis. Arch Neurol Psychiatry 44: 841

Neophytides AN, Di Chiro G, Barron SA, Chase TN (1979) Computed axial tomography in Huntington's disease and persons 'at risk' for Huntington's disease. In: Chase TN, Wexler NS, Barbeau A (eds) Advances in neurology, vol 23. Raven Press, New York, pp 185-192

Novom S, Danna S, Goldberg MA (1976) Intention myoclonus in Huntington's disease. Bull Los Angeles Neurol Soc 41(2): 82-84

Oepen H (1963) Paroxysmale Störungen bei der Huntingtonschen Chorea. Arch Psychiatr Nervenkr 204: 245-261

Oliver JE (1970) Huntington's chorea in Northamptonshire. Br J Psychiatry 166: 241-253

Panse F (1942) Die Erbchorea: eine Klinische-genetische Studie. Samml Psychiat Neurol Einzeldarst 18. Thieme, Leipzig

Patzig F (1935) Vererbung von Bewegungsstorungen. Z Abstamm Vererb-Lehre 70: 476-484

Paulson GW (1979) Diagnosis of Huntington's disease. In: Chase TN, Wexler NS, Barbeau A (eds) Advances in neurology, vol 23. Raven Press, New York, pp 177-184

Pearce J, Miller E (1973) In: Clinical aspects of dementia. Ballière Tindall, London, pp 16-66

Petit H, Milbled G (1973) Anomalies of conjugate ocular movements in Huntington's chorea: application to early detection. In: Barbeau A, Chase TN, Paulson GW (eds) Huntington's chorea, 1872-1972. Raven Press, New York, pp 287-294

Purdon Martin J (1967) The basal ganglia and posture. Pitman Medical, Tunbridge Wells

Refsum SB (1938) Et tilfelle ar Huntington's chorea med utpregede vegetative symptome. Nor Mag Laegevid 99: 1201-1218

Rosenberg RN, Thomas L, Baskin F, Kirkpatrick J, Bay C, Nyhan WL (1979) Joseph disease: protein patterns in fibroblasts and brain. Neurology (Minneap) 29: 917-926

Sandberg PR, Fibiger HC (1979) Body weight, feeding and drinking behaviour in rats with kainic acid-induced lesions of striatal neurones, with a note on body weight symptomatology in Huntington's disease. Exp Neurol 66: 444-466

Scott DF, Heathfield KW, Toone B, Margerison JH (1972) The EEG in Huntington's chorea: a clinical and neuropathological study. J Neurol Neurosurg Psychiatry 35: 97-102

Shoulson I (1977) Clinical care of the patient and family with Huntington's disease. In: Report to Commission for the Control of Huntington's Disease, vol II, Technical Report. US Government Printing Office, Washington, DC, pp 421-442

Shoulson I, Fahn S (1979) Huntington's disease: clinical care and evaluation. Neurology 29: 1-3

Starr A (1967) A disorder of rapid eye movements in Huntington's chorea. Brain 90: 545-564

Stevens DL (1973) The classification of variants of Huntington's chorea. In: Barbeau A, Chase TN, Paulson GW (eds) Huntington's chorea, 1872-1972. Raven Press, New York, pp 51-64

Stevens DL (1976) Huntington's chorea: a demographic, genetic and clinical study. MD thesis, University of London, pp 1-338

Stevens DL, Parsonage MJ (1969) Mutation in Huntington's chorea. J Neurol Neurosurg Psychiatry 32: 140-143

Terrence CF, Delaney JF, Alberts MC (1977) Computed tomography for Huntington's disease. Neuroradiology 13: 173-175

Waters CO (1842) Letter dated 5 May 1841. In: Dunglison R, Practice of medicine, 1st ed. Lee and Blanchard, Philadelphia, p 245

Weingartner H, Caine ED, Ebert MH (1979) Encoding processes, learning and recall in Huntington's disease. In: Chase TN, Wexler NS, Barbeau A (eds) Advances in neurology, vol 23. Raven Press, New York, pp 215-226

Westphal C (1883) Über eine dem Bilde der Cerebrospinalen Grauen Degeneration ähnliche Erkrankung des zentralen Nervensystems ohne anatomischen Befund, nebst einigen Bemerkunger über paradoxe Kontraktion. Arch Psychiatr Nervenkr 14: 87-96; 767-773

Whittier JR (1967) Clinical aspects of Huntington's disease. In: Barbeau A, Brunette JR (eds) Progress in neurogenetics, vol I. Excerpta Medica, Amsterdam, pp 632-644

Wilson RS, Garron DC (1979) Cognitive and affective aspects of Huntington's disease. In: Chase TN, Wexler NS, Barbeau A (eds) Advances in neurology, vol 23. Raven Press, New York, pp 193-202

6 Neuropathology

Current concepts concerning the neuropathology of Huntington's chorea are briefly reviewed in this chapter. It is written for the practising clinician and readers requiring more detailed information are urged to consult the excellent reviews on this subject by Bruyn (1968, 1973; Bruyn et al. 1979), Dunlap (1927), Earle (1973), Hallervorden (1957), Klintworth (1973) and Forno and Norville (1979).

The pathology of Huntington's chorea remained misinterpreted for some time after Huntington's original description in 1872. Even though certain workers had recognised the relationship of chorea to neostriatal neuronal loss (Golgi 1874; Meynert 1877) this view was not accepted by the majority, who considered that the structural abnormalities in brains of patients dying from Huntington's chorea were consistent with a meningitis (Phelps 1892), an encephalitis (Clarke 1897; Damaye 1909; Facklam 1898; Wiedenhammer 1901) and an intracranial haemorrhage (Jones 1905). Even though Besta (1905) and others (Bonfigli 1908; Evans 1908) correctly noted cortical cell destruction they wrongly assumed that it was secondary to an inflammatory process.

The possible significance of the striatal destruction as a primary feature in the pathology of Huntington's chorea was revived after the classic papers of Anglade (1906) and Jelgersma (1908). Alzheimer reported in 1911 that the main features of this disorder were cortical and neostriatal neuronal loss and these findings were subsequently more frequently documented by other authors (Kisselbach 1914; Marie and Lhermitte 1914; Pfeiffer 1914) who, in addition, postulated that there might be some relationship between the choreiform movements and the basal ganglia destruction.

Despite the fact that this concept was reiterated repeatedly, controversy continued for some time. Finally, the situation was clarified by Dunlap (1927), who stated that the combination of changes in the corpus striatum and cerebral cortex was so characteristic as to permit a diagnosis of Huntington's chorea to be made, even without a clinical history.

Despite numerous reports of widespread changes in different parts of the central nervous system, the understanding of the pathology of Huntington's chorea remained essentially constant for approximately 40 years until new stimulus was provided by the application of different histochemical techniques and electron microscopy (Sluga-Gasser 1966) to the disorder. With these developments, there has been a move away from pure descriptive morphology towards attempts to examine and clarify the pathological processes which may antedate the previously well-recognised lesions in the cortex and neostriatum. The recent endeavours to clarify the neuropathology of Huntington's chorea have included cytometric studies and examination of the nuclear–nucleolar system in affected persons (Roizin et al. 1979). Quantitative investigations of the neuronal cell loss in various regions of the brain and the ratio between different cell types in similar areas

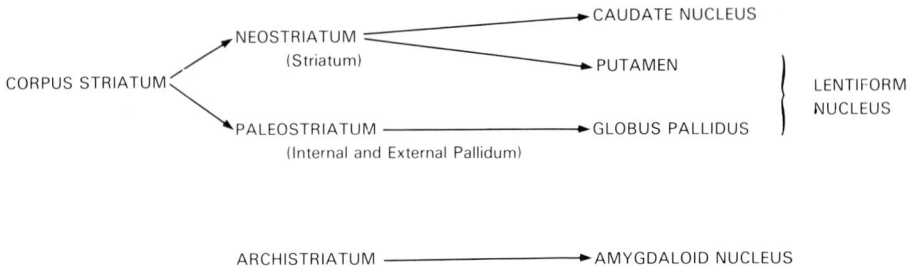

```
                                                        ┌──────────► CAUDATE NUCLEUS
                                    ╱ NEOSTRIATUM ◄──────┤
                                   ╱    (Striatum)        └──────────► PUTAMEN                    ⎫
                CORPUS STRIATUM ◄                                                                 ⎬  LENTIFORM
                                   ╲                                                              ⎭  NUCLEUS
                                    ╲ PALEOSTRIATUM ─────────────────► GLOBUS PALLIDUS
                                         (Internal and External Pallidum)


                          ARCHISTRIATUM ─────────────────────────────► AMYGDALOID NUCLEUS
```

Fig.6.1. A classification of the basal ganglia

may provide more rational explanations for the variation in clinical presentation in the disorder.

Confusion still exists with regard to the nomenclature of the basal ganglia and a currently accepted classification is shown in Fig. 6.1.

6.1. Gross Pathology

The meninges are often thickened and opaque and there is a general diminution in brain size, with a reduction in weight of between 150 and 450 g. This is primarily due to atrophy which particularly affects the cerebral cortex and the deep grey matter and, as a result, the gyri are narrowed and the sulci widened. Bruyn et al. (1979) estimate that there is a 20% reduction in cortical volume and this accounts for most of the brain weight loss. On coronal section of the brain the most striking abnormality is the bilateral atrophy of the caudate nucleus and putamen which results in dilatation of the lateral ventricles (Fig. 6.2). Atrophy of the globus pallidus may also be present, but to a mild degree. The cerebellum is usually not macroscopically affected.

6.2. Findings on Light Microscopy

6.2.1. Leptomeninges

The meninges are generally thickened and occasionally monocytic and lymphocytic infiltration are present.

6.2.2. Cerebral Cortex

Neuronal loss affects mainly the deeper regions of the cortex, particularly the third, fourth and fifth layers. However, cells in other levels may also show disturbance of architecture, with malalignment and distortion of cell shapes. There is occasional low-grade macrophage activity which occurs primarily around blood vessels (Tellez-Nagel et al. 1973). Lipopigment is generally abundant and an impression of astrocytic proliferation is gained, primarily due to tissue atrophy. Changes in subcortical white matter also occur, with shrinkage of cells and myelin loss.

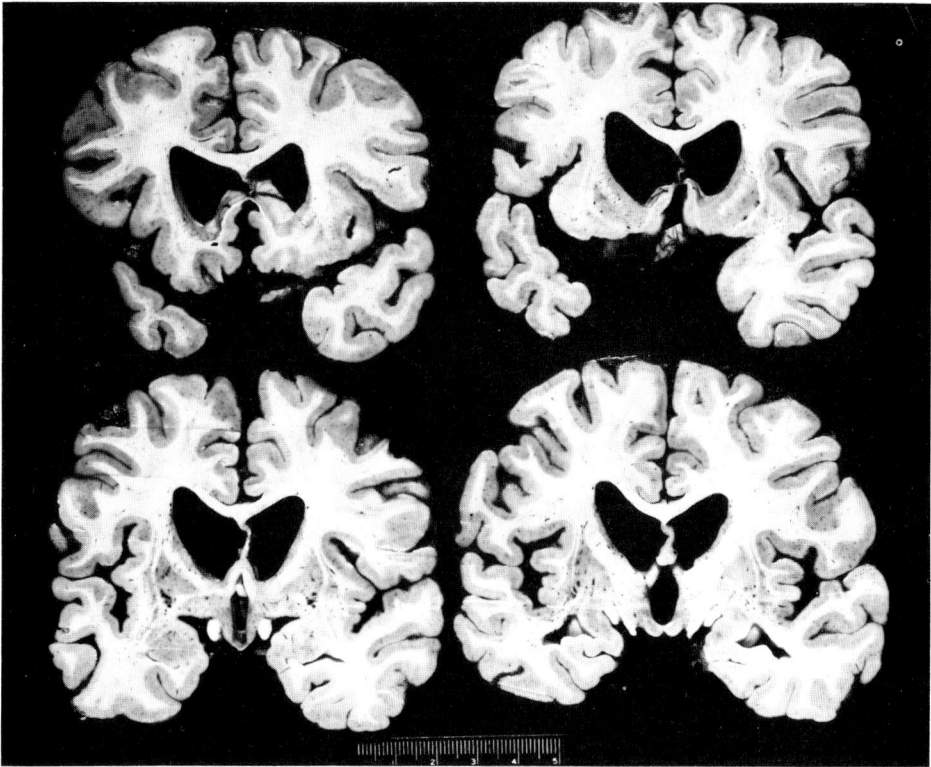

Fig.6.2. Coronal section of the brain of a patient with juvenile Huntington's chorea. The caudate nucleus is markedly flattened, the putamen is shrunken and the ventricular system is dilated. Mild cortical atrophy and widening of sulci are present. (By courtesy of F. Gilles, MD, and the Editor of *Neurology*)

6.2.3. The Caudate Nucleus and Putamen

The caudate nucleus and putamen show marked neuronal loss (Fig. 6.3), the small cells (Golgi type II) being affected almost four times as commonly as the large neurones. There is general consensus that iron is consistently but variably increased in the striatum and pallidum. Klintworth (1973) has questioned whether these iron deposits point to an underlying defect in iron metabolism. However, this is unlikely as increased iron deposit is certainly not specific to Huntington's chorea and may occur in brains of patients with Parkinson's disease, Pick's disease and senile dementia (Earle 1968).

It has generally been accepted (Bruyn 1968; Klintworth 1973) that a feature of the pathology of Huntington's chorea is an increased number of astrocytes and astrocytic processes (Fig. 6.4). However, this has recently been disputed by Lange and Thorner (unpublished), who demonstrated a 25% diminution of glial cells in the neostriatum. It is thought that apparent glial cell proliferation is probably due to the greater magnitude of neuronal cell loss with resultant relative glial cell condensation. Bruyn et al. (1979) have suggested that Huntington's chorea could possibly be regarded as a disease of disturbed neuronoglial relationship resulting from some failure of glial function, such as removal of a possible toxin, or inadequate nutritional supply. This new concept deserves further consideration and illustrates the value of accurate documentation of the pathological features in the unravelling of the pathophysiology of the disorder.

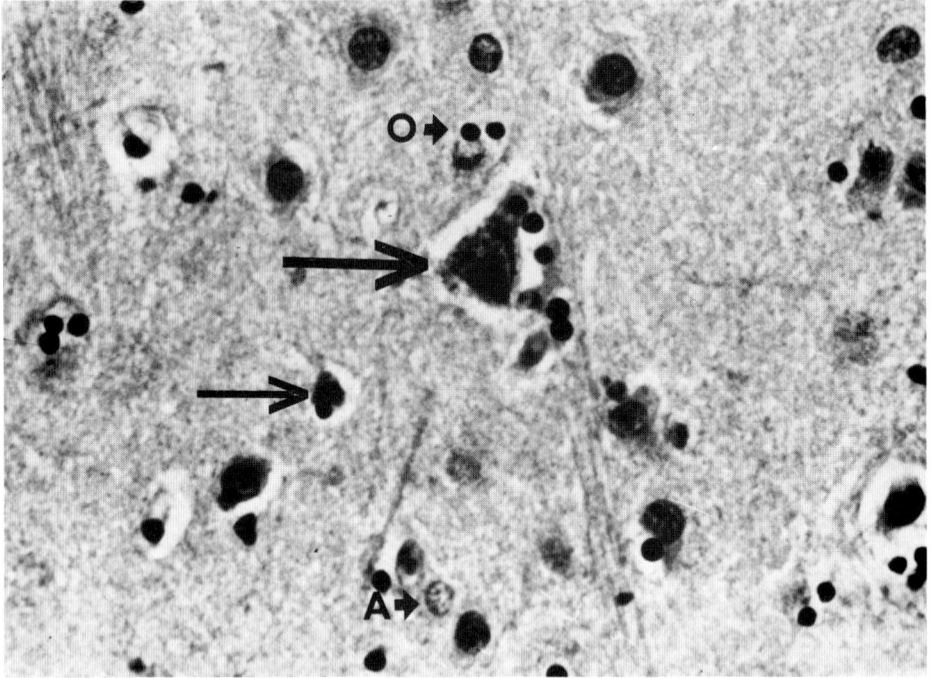

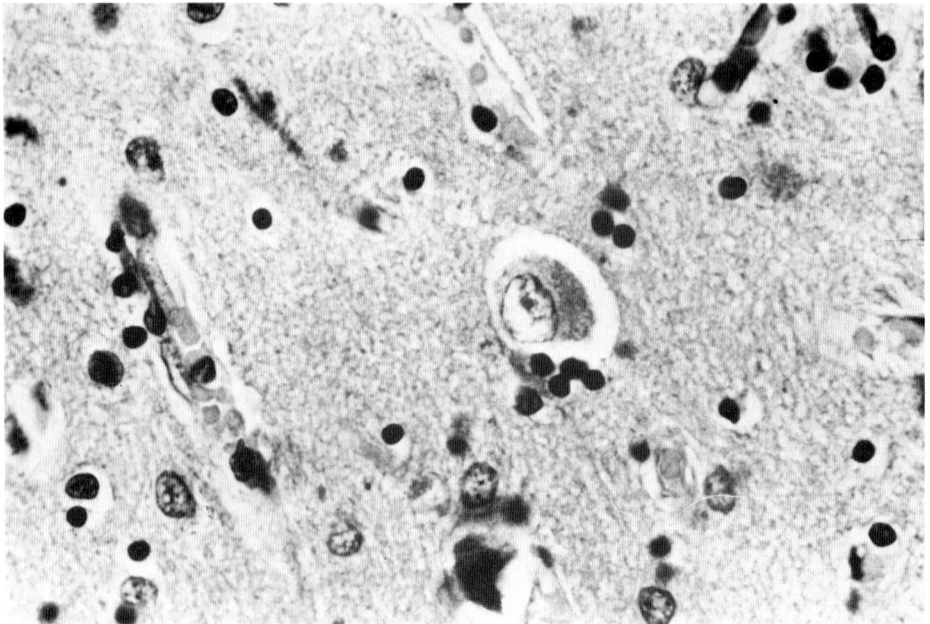

a

b

Fig.6.3.a Section from normal caudate nucleus showing large neurone (large arrow) and numerous small neurones (small arrow), astrocytes (A) and oligodendrocytes (O). H&E stain, ×400. (By courtesy of J. Morris, MD)

    b Caudate nucleus of an elderly woman with Huntington's chorea. Note the neuronal depletion and the relative predominance of astrocytes and oligodendrocytes. H&E stain, ×400. (By courtesy of T. Bird, MD)

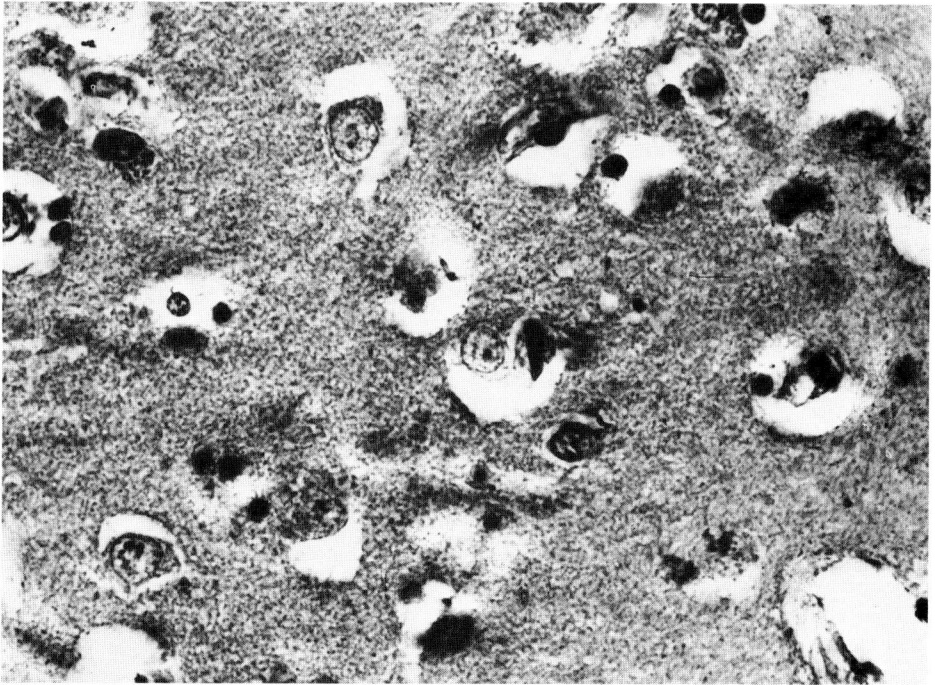

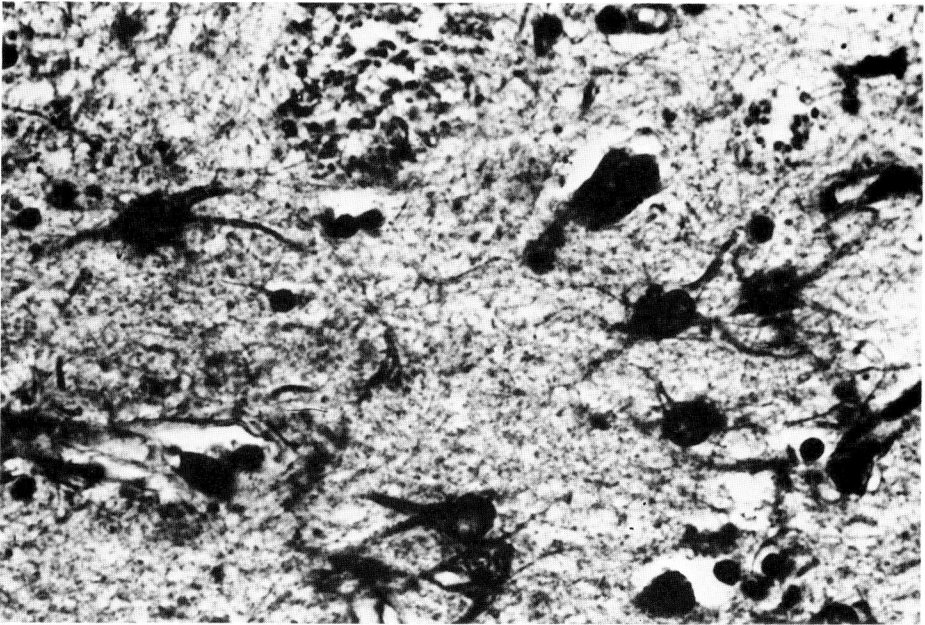

Fig.6.4.a Section from normal caudate nucleus. No astrocytic processes are visible. PTAH stain, ×400
  b Caudate nucleus of a patient with long-standing Huntington's chorea. Astrocytes and astrocytic processes are prominent. PTAH stain, ×400. (**a** and **b** by courtesy of J. Morris, MD)

6.2.4. Pathological Changes in Other Parts of the Nervous System

Mild abnormalities are frequently present in other parts of the central nervous system. The globus pallidus is commonly affected and may be reduced to almost half its original size consequent to neuronal loss, which particularly affects large cells. Cell loss is found less frequently in the subthalamic nucleus and substantia nigra. The latter area is often more darkly pigmented than normal, with marked atrophy of the zona reticulata compared with the zona compacta (Bird 1978). This is in marked contrast to the findings in Parkinson's disease, where the striatum generally appears normal whilst there is marked cellular degeneration in the pars compacta of the substantia nigra. Although some confusion initially existed with regard to the involvement of the thalamus, it is now clearly established that neuronal depopulation also occurs in this area (Dom et al. 1976; Forno and Jose 1973).

Cell loss in the supraoptic, ventromedial and lateral hypothalamic nuclei may partly explain the disturbed hypothalamic-pituitary function that has been reported in Huntington's chorea (Sect. 10.8).

The cerebellum is occasionally affected (particularly in juvenile patients) but the involvement is mild and usually comprises different degrees of Purkinje cell loss. Centres in the brain stem, including the superior-olivary, vagal and hypoglossal nuclei, are frequently involved, with some cell loss. The only pathological feature of note in the spinal cord is pallor, particularly affecting the anterolateral tracts.

None of these features is specific to Huntington's chorea. The regional distribution of these lesions, however, characterises the pathology of this disorder.

6.3. Ultrastructural Features

There have been remarkably few reports on the ultrastructural characteristics of the brain in Huntington's chorea and the electron microscopic studies which have been undertaken have focused primarily on the neostriatum and cortex. The landmarks in the description of the ultrastructure of Huntington's chorea are summarised in Table 6.1. Some of the main features will be discussed below.

Table 6.1. Some milestones in the description of the ultrastructural characteristics in Huntington's chorea

1966	Sluga-Gasser reported on the structure of the striatum of one patient
1973	Tellez-Nagel et al. documented the electron microscopic findings of cerebral biopsies of four persons with Huntington's chorea
1976	Roizin et al. suggested that the degenerative process may be related to a disorder of intracellular metabolism
1977	Bots reported the presence of lysosomal lipopigment in glial cells of the caudate nucleus, which differs from that seen in ageing
1978	Goebel et al. reported on the ultrastructural characteristics of the brain of a patient with juvenile Huntington's chorea, which conformed with findings in affected adults
1979	Roizin et al. studied the neuronal nuclear-cytoplasmic changes in 18 patients
1979	Forno and Norville reviewed the ultrastructure of the neostriatum in Huntington's chorea and compared the findings with those seen in Parkinson's disease

6.3.1. Cerebral Cortex

Tellez-Nagel et al. (1973) and Roizin et al. (1976) have reported a range of ultrastructural findings which include:

1) the accumulation of lipofuscin in neurones;

2) an increase in the smooth endoplasmic reticulum and vesicles associated with the Golgi complexes;

3) mitochondrial changes, including disrupted and fewer cristae;

4) degeneration of presynaptic endings, whilst the postsynaptic elements and synaptic junctions are structurally normal;

5) axonal degeneration;

6) proliferation and hypertrophy of the astrocytes.

Many of these abnormalities may be found in other diseases and in the ageing brain, and the significance of these findings in Huntington's chorea is at present poorly understood.

6.3.2. The Striatum

In general, electron microscopic studies of the neostriatum have defined similar abnormalities to those found in the cerebral cortex (Forno and Norville 1979). There is loss of neurones and general rarefaction of the neuropil, accompanied by an increase in

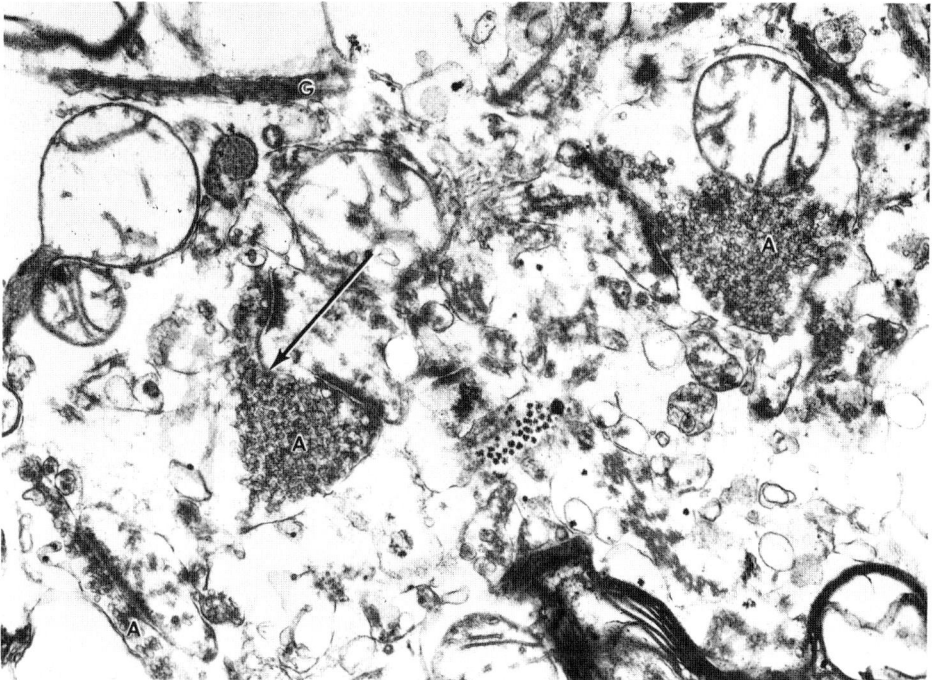

Fig.6.5. Electron micrograph of a 54-year-old man with long-standing Huntington's chorea. General rarefaction of the neuropil and increase in bundles of glial elements (G) are seen. Several preserved axons (A) with synaptic vesicles are present. An asymmetrical interrupted synapse is indicated by the arrow. ×21 000 (By courtesy of L. Forno, MD)

the bundles of glial elements in astrocytic processes and a normal or relatively increased number of oligodendroglial cells (Fig. 6.5). Bots (1977) established that there is a deposition of lipopigment in glial cells which is clearly distinguishable from that seen in senescence, and this finding may further implicate glial cell involvement in the pathogenesis of the disorder.

Comparative electron microscopic studies on brains of controls and those from patients with other chronic neurological diseases are needed to substantiate and augment the results of these earlier reports. Furthermore, other areas of the brain, including the substantia nigra, hypothalamus and cerebellum, which have revealed some abnormalities under light microscopy, warrant electron microscopic investigation.

6.4. The Importance of the Post-mortem Examination

A surprising result of the autopsy studies of patients thought to have died from Huntington's chorea is that a relatively high proportion of these persons were, in fact, suffering from some other neurological condition. Corsellis (1976) and Bird (1979) have reported that approximately 7% of their series were erroneously diagnosed in this way. The condition most commonly misdiagnosed as Huntington's chorea was found to be Alzheimer's disease (Table 5.8). This is particularly serious in view of the different genetic and social implications of these two diseases. Equally serious and perhaps more frequent is the failure of recognition of Huntington's chorea until autopsy. In this context, it is relevant that Bird (1979) has reported that two patients with diagnosed schizophrenia in whom post-mortem examination was undertaken for other reasons were found to have pathological changes consistent with Huntington's chorea.

In the light of these findings and in view of the relatively high frequency of clinical misdiagnosis and absence of any definitive diagnostic procedure, it is reasonable to suggest that all persons dying of Huntington's chorea deserve post-mortem examination. The establishment of a precise diagnosis is of great practical application and facilitates appropriate genetic counselling. There is also great merit in Bird's (1979) statement that an autopsy is justified for every member of an afflicted family, be they affected or unaffected. This would allow confirmation of the suspected diagnosis in the affected person and also permit examination of the brain of those clinically normal at-risk persons who have died from other causes. Thereby it may be possible to establish whether the pathological lesions antedate the clinical features. Furthermore, examination of the brains of mildly affected individuals may give additional clarification as to the temporal sequence of development of the pathological lesions in Huntington's chorea.

Whilst most patients with advanced disease have the typical and well-described pathological features of this disorder, a most surprising and unusual finding has been the remarkable preservation of neurones in some patients with gross clinical symptomatology (Earle 1973; Oepen 1973). In other words, there would seem to be no consistent relationship between the extent of cerebral or striatal atrophy and the duration or presentation of the clinical illness. Further studies are clearly needed to examine more closely the neuropathological correlation between structure and function in this disease.

A possible way of investigating this problem other than by post-mortem examination is by brain biopsy in vivo. There is much dispute concerning the ethics of this technique, in view of its possible but uncommon medical complications on the one hand and its uncertain use in terms of diagnosis on the other. The advent of a diagnostic test for Huntington's chorea would largely obviate the need for cerebral biopsy.

The pathological substrate of chorea is generally thought to be the destruction of the caudate nucleus. The substantia nigra (Bielschowsky 1922; Schroeder 1931), the putamen

(Denny Brown 1962) and globus pallidus (Byers et al. 1973) have all been suggested as the anatomic substrates of rigidity, but the correlative neuropathological basis of this sign remains uncertain.

The more frequent involvement of the cerebellum in juvenile patients, characterised by Purkinje cell loss (Klintworth 1973), is a feasible explanation for the presence of cerebellar signs in some persons with this form of the disorder.

Huntington's chorea shares with other causes of dementia, such as Alzheimer's, Creutzfeldt-Jacob and Pick's disease, the pathological characteristics of selectivity, progression and clinical-pathological variability. However, in contrast to the neuro-fibrillary tangles of Alzheimer's disease or the Levy body which is a feature of Parkinson's disease, there is at present no distinguishing cytological alteration in Huntington's chorea.

The nucleus accumbens is the most anterior portion of the caudate nucleus and its involvement in Huntington's chorea is currently attracting particular attention. This area of the striatum is relatively spared from neuronal cell loss until the disease is very advanced (Forno and Jose 1973) and may be more likely to show early specific cell damage in the disorder. Bruyn (1980) and his co-workers have been studying this area of the caudate nucleus but at the present time no results have been published.

A major problem in the elucidation of the primary pathological features of Huntington's chorea is the determination of the effects of other unrelated factors, such as the consequence of prolonged neuroleptic administration, on brain structure. In this context, it is relevant that there is at present no readily available technique for estimating neuroleptic concentration in post-mortem brains.

6.5. The Brain and Tissue Bank

The aim of a post-mortem tissue bank is to collect and preserve brain and other tissue for investigators who need such specimens for research. This is particularly important in Huntington's chorea, where the pathophysiology and basic defect have not yet been elucidated. Since the development in 1972 of a resource centre of this type in Cambridge, England, some progress has been made in the understanding of the neurochemistry of Huntington's chorea, with particular reference to the changes in the different neurotransmitters and their enzymes. A similar bank has now been established in Belmont, Mass., specifically for the collection of tissues from patients dying of Huntington's chorea in the USA.

The success of these centres depends on cooperation between affected families, managing physicians and the pathologists who perform the post-mortems. Families afflicted with Huntington's chorea are often anxious to collaborate with researchers in this way, in the hope that this may lead to a furthering of knowledge about the disorder.

Optimally, physicians can tactfully explain the need for post-mortem examination and obtain permission from the family prior to the affected person's death. This is most important, as it may diminish the interval between death and post-mortem and will thus possibly improve the quality of tissues received in the brain bank. This is turn will allow inferences to be made about the neurochemistry of Huntington's chorea with the knowledge that the changes are unlikely to be due to post-mortem artefact.

A full protocol for the collection of brains and other non-cerebral tissues such as kidneys and livers is available from the offices of the lay organisations in the United States and England, and from the brain bank centres in those countries, whose addresses are to be found in Appendix 5. These brain banks operate at an international level and arrangements can be made for families and physicians outside the United States and England to contribute to this resource centre. Similarly, workers in other countries who are involved in research on Huntington's chorea and need brain tissue may approach the centre for this purpose.

References

Anglade M (1906) Une autopsie de chorée de Huntington. Gaz Hebd Sci Med Bordeaux 27:89

Alzheimer A (1911) Über die anatomische Grundlage der Huntington'schen Chorea und der choreatischen Bewegungen überhaupt. Neurol Zbl 30: 891-892

Besta C (1905) Un caso di corea di Huntington con reporto anatomopatologico. Riv Sper Freniat 31: 205-231

Bielschowsky M (1922) Weitere Bemerkungen zur normalen und pathologischen Histologie des Striaten Systems. J Psychol Neurol (Leipzig) 27: 233-288

Bird ED (1978) The brain in Huntington's chorea. Editorial. Psychol Med 8: 357-360

Bird ED (1979) Communication to Eighth International Workshop on Huntington's Chorea, Oxford, England

Bonfigli R (1906) Progressive chronic chorea: a clinical and anatomopathological study. J Ment Pathol 8: 63-73

Bots G (1977) Ultrastructural changes in cortex of Huntington patients. Communication to Seventh International Workshop on Huntington's Chorea, Leiden

Bruyn GW (1968) Huntington's chorea: historical, clinical and laboratory synopsis. In: Vinken PJ, Bruyn GW (eds) Handbook of clinical neurology, vol VI. North-Holland, Amsterdam, pp 298-378

Bruyn GW (1973) Neuropathological changes in Huntington's chorea. In: Barbeau A, Chase TN, Paulson GW (eds) Huntington's chorea, 1872-1972. Raven Press, New York, pp 399-403

Bruyn G (1980) Lecture to South African Neurology Symposium, Johannesburg

Bruyn GW, Bots TAM, Dom R (1979) Huntington's chorea: current neuropathological status. In: Chase TN, Wexler NS, Barbeau A (eds) Advances in neurology, vol 23. Raven Press, New York, pp 83-93

Byers RK, Gilles FH, Fung C (1973) Huntington's disease in children. Neurology 6: 561-569

Clarke JM (1897) On Huntington's chorea. Brain 20: 22-34

Corsellis JAN (1976) Ageing and the dementias. In: Blackwood W, Corsellis J (eds) Greenfield's neuropathology. Edward Arnold, London, p 822

Damaye H (1909) Autopsie de deux cas de chorée chronique avec troubles mentaux à la période démentielle. Rev Psychiatr Psychol Exp (Paris) 13: 621-629

Denny Brown D (1962) The basal ganglion and their relation to disorders of movement. Oxford University Press, London, p 25

Dom R, Baro F, Brucher JM (1973) A cytometric study of the putamen in different types of Huntington's chorea. In: Barbeau A, Chase TN, Paulson CW (eds) Huntington's chorea, 1872-1972. Raven Press, New York, pp 369-383

Dom R, Malford M, Baro F (1976) Neuropathology of Huntington's chorea: cytometric studies of the venterobasal complex of the thalamus. Neurology 26: 64-68

Dunlap CB (1927) Pathologic changes in Huntington's chorea with special reference to the corpus striatum. Arch Neurol Psychiatry 18: 867-943

Earle KM (1968) Studies on Parkinson's disease, including X-ray, fluorescent spectrography of formalin-fixed brain tissue. J Neuropathol Exp Neurol 27: 1-14

Earle KM (1973) Pathology and experimental models of Huntington's chorea. In: Barbeau A, Chase TN, Paulson GW (eds) Huntington's chorea, 1872-1972. Raven Press, New York, pp 439-452

Evans JJW (1908) Observations on a case of Huntington's chorea. Lancet II: 940

Facklam FC (1898) Beiträge zur Lehre vom Wesen der Huntington'schen Chorea. Arch Psychiatr Nervenkr 30: 137-204

Forno LS, Jose C (1973) Huntington's chorea: a pathological study. In: Barbeau A, Chase TN, Paulson GW (eds) Huntington's chorea, 1872-1972. Raven Press, New York, pp 453-470

Forno LS, Norville RL (1979) Ultrastructure of the neostratum in Huntington's and Parkinson's disease. In: Chase TN, Wexler NS, Barbeau A (eds) Advances in neurology, vol 23. Raven Press, New York, pp 123-135

Goebel HH, Peperitz R, Scholz W, Igbal K, Tellez-Nagel I (1978) Juvenile Huntington's chorea: clinical, ultrastructural and biochemical studies. Neurology (Minneap) 28: 23-31

Golgi C (1874) Sulla alterazioni degli organi centrali nervosi in uno caso di corea gesticulatoria assoziata ad alienazione mentale. Riv Clin Bologna 4: 361

Hallervorden J (1957) Huntingtonsche Chorea (chorea chronica progressiva hereditaria). In: Lubarsch O, Henke F, Rossle R, Uhlinger E (eds) Erkrankungen des Nervensystems, Handbuch Spez. Pathol. Anat. Histologie, vol 13/1A. Springer, Berlin, pp 793-822

Jelgersma G (1908) Die anatomische Veränderungen bei Paralysis agitans und chronischer Chorea. Verh Ges Dtsch Naturf Ärzt 2(2): 383-388

Jones R (1905) Huntington's chorea and dementia. Lancet 83 (II): 1831-1832

Kisselbach G. (1914) Anatomischer Befund eines Falles von Huntingtonscher Chorea. Monatschr Psychiatr Neurol 35: 525-543

Klintworth GK (1973) Huntington's chorea: morphologic contribution to a century. In: Barbeau A, Chase TN, Paulson GW (eds) Huntington's chorea, 1872-1972. Raven Press, New York, pp 543-549

Marie P, Lhermitte J (1914) Lesions de la chorée chronique progressive. La dégénération atrophique cortico-striée. Ann Med (Paris) 1: 18-48

Meynert T (1877) Discussion to Fritsch. Psychiatr Club 7: 47

Oepen H (1973) Ontogenetic and phylogenetic aspects in the neuropathology of Huntington's chorea. In: Barbeau A, Chase TN, Paulson GW (eds) Huntington's chorea, 1872-1972. Raven Press, New York, pp 425-427

Pfeiffer JAF (1914) A case of chronic progressive chorea with anatomical study. Am J Insan 71: 581-602

Phelps RM (1892) A new consideration of hereditary chorea. J Nerv Ment Dis 19: 765-776

Roizin L, Kaufman MA, Willson N, Stellar S, Liu JC (1976) Neuropathologic observations on Huntington's chorea. In: Zimmerman WM (ed) Progress in neuropathology, vol III. Grune & Stratton, New York, pp 447-488

Roizin L, Stellar S, Liu JC (1979) Neuronal nuclear cytoplasmic changes in Huntington's chorea: electron microscope investigations. In: Chase TN, Wexler NS, Barbeau A (eds) Advances in neurology, vol 23. Raven Press, New York, pp 95-122

Schroeder K (1931) Zur Klinik und Pathologie der Huntington'schen Krankheit. J Psychol Neurol 43: 183-201

Sluga-Gasser E (1966) Zur Ultrastruktur des Striatums. Vorlaufige Ergebnisse einer cerebralen Biopsie. Wien Z Nervenheilk 23: 17-35

Tellez-Nagel I, Johnsen AB, Terry RD (1973) Ultrastructural and histochemical study of cerebral biopsies in Huntington's chorea. In: Barbeau A, Chase TN, Paulson GW (eds) Huntington's chorea, 1872-1972. Raven Press, New York, pp 387-398

Weidenhammer W (1901) Zur pathologischen Anatomie der Huntingtonschen Chorea. Neurol Zentralbl 20: 1161-1162

7 Genetics

Huntington's description of the hereditary characteristics of this disease in his famous essay of 1872 was remarkably accurate. He described that "when either or both parents have shown manifestations of the disease, one or more of the offspring invariably suffer from the disease. It never skips a generation to again manifest itself in another. Once having yielded its claims it never regains them." Huntington's description based on his own empiric observations occurred only 6 years after Mendel's classic paper. The interpretation of the familial aggregation of Huntington's chorea in Mendelian terms, however, had to wait 36 years until Jelliffe (1908) drew attention to the fact that the mode of inheritance of this disease "shows a very close approximation to the results of Mendelian crossing".

Since that time it has generally been accepted that Huntington's chorea is inherited as an autosomal dominant trait. This implies that the gene is transmitted by parents of either sex to children of either sex. Each child of an affected parent has an even chance of inheriting the gene (Fig. 7.1). This does not mean, as is commonly believed, that half the offspring of a given sibship will develop the disease but rather that each descendant of an affected person has a 50% chance of receiving the abnormal gene, irrespective of the number of sibs who have already shown signs of the disorder. All heterozygotes for Huntington's chorea will show signs and symptoms if they live long enough. In other words, incomplete penetrance, as seen in other autosomal dominant conditions, does not occur. Penetrance is a statistical concept and refers to the fraction of those affected who have the given abnormal gene and manifest the expected, specified phenotype.

If an individual does not carry the gene for Huntington's chorea, the gene cannot be transmitted to his or her progeny. An impression of incomplete penetrance or skipping of generations may occur if a child of a clinically normal parent, but with an affected grandparent, develops features of the disease prior to onset in the parent. This is a definitive indication that this parent is, in fact, an unsuspected heterozygote for Huntington's chorea and will develop features of the disorder. No physical characteristics of either the parents or the offspring will alter the inheritance of Huntington's chorea.

'Anticipation' refers to the presentation of recognisable clinical features of a disease at progressively earlier ages in successive generations. This phenomenon has been thought to occur in different autosomal dominant disorders, including myotonic dystrophy (Penrose 1948) and Huntington's chorea (Heilbronner 1903). However, the concept has fallen into disrepute with regard to the latter disease as it is now thought to reflect a statistical artefact rather than the true situation.

Disorders of simple Mendelian inheritance are not usually associated with chromosomal abnormalities. Huntington's chorea has nevertheless been subjected to scrutiny

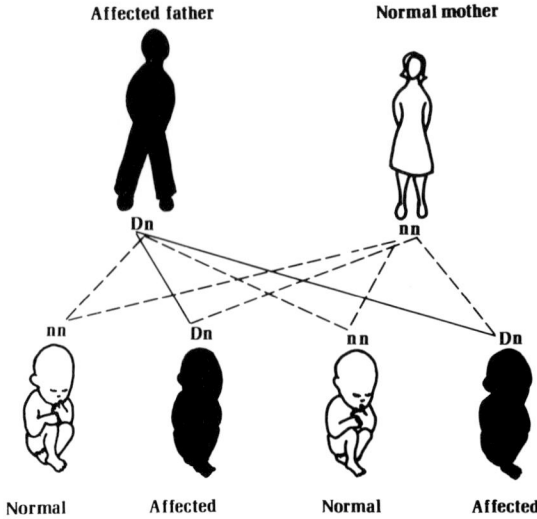

Affected father   **Normal mother**

Dn   nn

nn   Dn   nn   Dn

Normal   **Affected**   **Normal**   **Affected**

Fig.7.1. Mechanism of autosomal dominant inheritance. The gene is transmitted from generation to generation by parents of either sex to children of either sex

according to the cytogenetic investigative techniques developed in recent years, but no abnormalities of this nature have been found (Delhanty 1979).

In spite of the disease's long history on the one hand and the advances in research concerning molecular biology on the other, there has been remarkably little progress in the understanding of the hereditary nature of Huntington's chorea. For a long time the mode of inheritance was thought to conform to accepted principles of Mendelian genetics. Recently, however, certain anomalous findings have been reported which suggest that the genetics of Huntington's chorea is more complex than initially suspected.

In this chapter a review of established concepts of the genetics of the disorder will be presented and some of the unresolved issues will be discussed.

7.1. Mutations

The term mutation was introduced by de Vries at the turn of this century as a result of his observations of the genetic changes that occurred in his study of a particular plant species. Since that time it has been accepted that alterations in the carriers of genetic information may occur and that these changes may be transmitted to future generations.

The gene for Huntington's chorea may also arise spontaneously in this way but this is generally accepted to be a rare occurrence. At the present time there are few reports of mutations in Huntington's chorea which fulfil the criteria suggested by Stevens and Parsonage (1969). These are:

1) The disease must be Huntington's chorea and descend for more than one generation from the mutant.

2) The parents of the mutant must have died after the age of 70 years and must have been symptom-free.

3) The apparent mutant must be the child of his alleged parents and not the illegitimate issue of an unknown choreic.

Bell (1934) in her comprehensive publication gave six examples from her collection of pedigrees from the literature of persons with typical chorea whose parents reached old age

Table 7.1.   Mutation rate of the gene for Huntington's chorea (mutation rate/locus/generation)

Author(s)	Direct estimate	No. of mutants/10^6 gametes	Indirect estimate	No. of mutants/10^6 gametes
Kishimoto et al. (1957)	—	—	6.7×10^{-7}	0.67
Reed and Neel (1959)	5.4×10^{-6}	5.4	9.6×10^{-6}	9.6
Mattsson (1974)	5.0×10^{-6}	5.0	0.8×10^{-6}	0.8
Stevens (1976)	4.0×10^{-6} 4.22×10^{-7}	4.0 } 0.42 }	7×10^{-7}	0.7
Hayden (1979)	1.3×10^{-7}	0.13		

Table 7.2.   Selected mutation rates for some autosomal dominant conditions

Condition	Location	Mutation rate	No. of mutants/10^6 gametes
Achondroplasia	Denmark	1×10^{-5}	10
Dystrophia myotonica	Northern Ireland	8×10^{-6}	8
	Switzerland	1.1×10^{-5}	11
Osteogenesis imperfecta	Sweden	$0.7\text{-}1.3 \times 10^{-5}$	7-13
	Germany	1.0×10^{-5}	10
Tuberous sclerosis	Oxford, England	1.05×10^{-5}	10.5
	China	6×10^{-6}	6
Neurofibromatosis	Michigan, USA	1×10^{-4}	100
	Moscow	$4.4\text{-}4.9 \times 10^{-5}$	44-49
Intestinal polyposis	Michigan, USA	1.3×10^{-5}	13
Polycystic disease of the kidney	Denmark	$6.5\text{-}12 \times 10^{-5}$	65-120

From Vogel and Motulsky (1979).

(over 70) without showing any symptoms of the disease and who transmitted the gene to their offspring according to the laws of autosomal dominant inheritance (Geratovitsch 1927; Kronthal and Kalischer 1895; Osler 1893; Schulz 1898; Seip 1928; Smith 1898). Whilst paternity was not proven in any of these instances, it is likely that the affected person represented a spontaneous mutation for Huntington's chorea. Stevens and Parsonage (1969) have also reported another possible mutation for this disease. It is, of course, obvious that proof of legitimacy is impossible once the parents concerned are dead, which is generally the situation once it becomes clear that the disease is developing in their grandchildren. However, one such study has been conducted by Chiu and Brackenridge (1976), who used 25 genetic markers to prove parentage of a suspected mutant for the disorder.

There are two methods for estimation of the mutation rate and these will be further described in Appendix 2. Results of prior estimates made by both methods are shown in

Table 7.1; the number of mutants for Huntington's chorea per million gametes varies between 0.13 and 9.6 with a mean of 2.9. Comparison with some other autosomal dominant conditions shows that the occurrence of spontaneous mutations in Huntington's chorea is indeed rare and amongst the lowest for any similarly inherited condition (Table 7.2).

7.2. Heterozygote Frequency

The heterozygote frequency of Huntington's chorea refers to the number of individuals in a given population who carry the gene for the disease, such persons being either clinically affected or asymptomatic. This parameter gives a more accurate representation of the burden of this disease to a given community than prevalence rates, as it includes both those already affected and those asymptomatic individuals who will eventually develop the disease. As a result, rational planning for the provision of health care facilities can take place. Determination of the heterozygote frequency can also be used as an indication of the efficiency of any programme designed to decrease the frequency of the gene for Huntington's chorea.

Numerous estimates of the heterozygote frequency have been made and these are shown in Table 7.3. Different methods have been used to calculate this value, the most simple and efficient being that used by Reed et al. (1958) and Stevens (1973); this is described in Appendix 3.

Table 7.3.   Estimates of heterozygote frequency

Author(s)	Location	Frequency	Population ratio
Pearson et al. (1955)	Minnesota, USA	21.23×10^{-5}	1 : 4 710
Reed et al. (1958)	Michigan, USA	10.1×10^{-5}	1 : 9 906
Stevens (1973)	Leeds, England	10.77×10^{-5}	1 : 9 285
Shokeir (1975)	Manitoba, Canada	23.3×10^{-5}	1 : 4 291
Harper et al. (1979)	South Wales	20.2×10^{-5}	1 : 4 950
Hayden and Beighton (1981)	South Africa	6.7×10^{-5}	1 : 14 925

7.3. The Homozygous Form

According to classic Mendelian principles the offspring of two parents heterozygous for Huntington's chorea would have a 50% chance of being heterozygous, a 25% chance of being homozygous for the gene and a 25% possibility of not inheriting the abnormal gene (Fig. 7.2).

Matings between two people affected with Huntington's chorea are extremely rare and only two prior occurrences of this nature have been well documented. In 1936 Hindringer reported the marriage of two cousins, both of whom developed the disease. Of their offspring, three had clinically typical Huntington's chorea, one died in infancy of unknown cause and the remaining offspring was unaffected at the age of 36.

Eldridge et al. (1973) reviewed Hindringer's patients and reported a second kindred in which a consanguineous union took place between two people who both developed

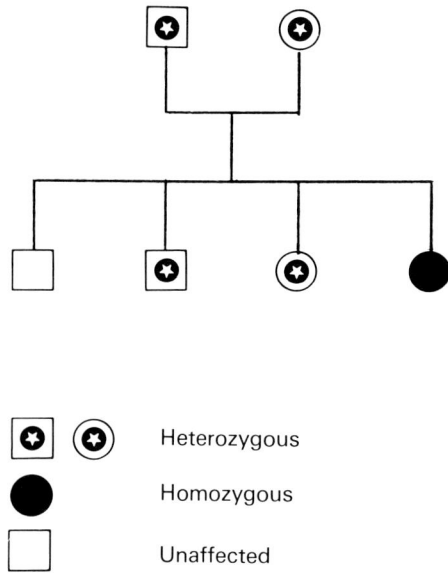

Fig.7.2. The expected mode of inheritance in the offspring of two parents heterozygous for Huntington's chorea

Huntington's chorea. They produced three children, only one of whom, the youngest, was typically affected. The other two offspring had an early demise. The eldest was a young man who had no mental or physical disorder and perished during combat in the Second World War, whilst the middle child was a term infant girl who died of unknown cause a few days after birth.

There is no way at present of knowing whether offspring of affected couples are heterozygous or homozygous for the Huntington's chorea gene. There were no unusual clinical or laboratory features in the abovementioned patients. It is, however, noteworthy that in both kindreds described there was an infant who, without obvious cause, died within a few days of birth. Whether these offspring were homozygous for the gene of Huntington's chorea is uncertain but this situation could possibly be evidence for the lethal effect of the 'double dose' of the abnormal gene.

There is evidence from the study of other autosomal dominant disorders, such as hereditary haemorrhagic telangiectasia (Snyder and Doan 1944) and familial hyper-cholesteraemia (Frederickson et al. 1978) that the homozygous form induces a more severe manifestation of the disease than that seen in the heterozygote, and it is very feasible that a similar situation occurs in Huntington's chorea. For this reason, persons homozygous for Huntington's chorea may be ideal subjects for investigation of the fundamental biochemical defect in the disease.

There is a particularly high prevalence of Huntington's chorea in the state of Zulia in Venezuela (Fig. 7.3). An interesting and unusual characteristic of this community is that 30% of the patients had two parents with the same illness (Avila-Giron 1973). This extraordinary situation has arisen as a result of the aggregation of affected persons in a relatively isolated area of the shores of Lake Maracaibo, secondary to their social ostracisation. The inevitable consequence has been intermarriage. This population offers a unique research opportunity for investigation of the characteristics of the homozygous form of Huntington's chorea, which could ultimately lead to more rapid identification of the primary defect in this disease.

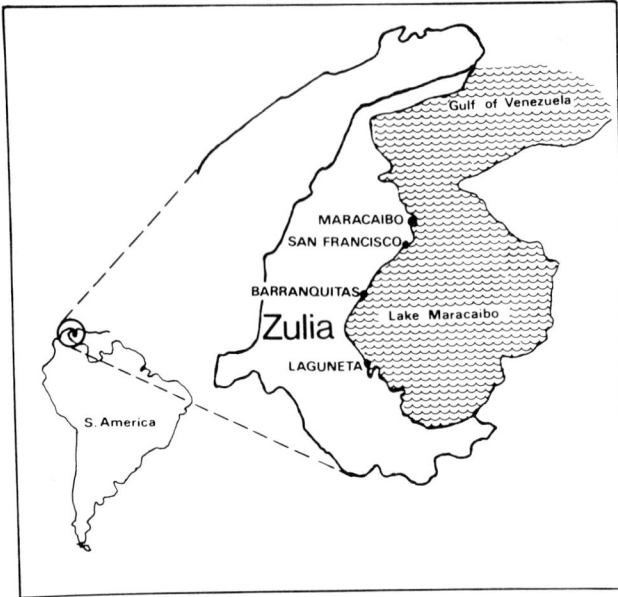

Fig.7.3. The state of Zulia in Venezuela, showing the towns of particularly high aggregation of patients with Huntington's chorea. (Adapted from the Report of the Commission to Control Huntington's Disease and its Consequences in the USA)

7.4. Heterogeneity

A common phenomenon in clinical genetics is that similar phenotypes are caused by more than one genotype. The discrimination of genetic entities in this way has been a target of much research effort in medical genetics in recent years, and Huntington's chorea has now also come under critical scrutiny.

At present it is still generally accepted that Huntington's chorea is caused by a single abnormal gene. Arguments in favour of heterogeneity have centred around the greater variation in clinical presentation of affected individuals between different families compared with that observed within kindreds. Age at onset and age at death data were analysed by Wallace and Hall (1972) and Went et al. (1975) and revealed greater intra-familial than interfamilial similarity (Wallace and Hall 1972). These findings were interpreted as supporting the existence of varying genetic forms of Huntington's chorea, possibly allelic in nature, which result in different age at onset and varying clinical features. But it has recently been shown that statistical significance ($P < 0.05$) could be reached for differences in age at onset between the two sibships of a single large Afrikaner kindred (J. Davidson and M. Hayden 1980, unpublished work), even though the disease was presumably caused by the same gene. In other words, finding statistical significance in relation to differences in clinical and natural history data between families is not, per se, definitive evidence of heterogeneity and may rather reflect intrafamilial similarities in genetic background. The possible causes of the marked variation in clinical presentations within one family are unknown.

The determination of the existence of heterogeneity in Huntington's chorea would have widespread implications. It could be the underlying reason for the differing clinical and biochemical responses in this disease, and could also be used as an important aid in rationalisation of therapy, if varying genetic forms responded to different medications.

A definitive method for establishing the presence of multiple genes in this disorder is by gene mapping. Genetic linkage is the most efficient method available for the chromosomal assignment of the locus of the abnormal gene. Thus far, the search for the locus has been unsuccessful; details of the different investigations that have been conducted will be presented in Chap. 10.

7.5. Unusual Aspects of the Genetics of Juvenile Huntington's Chorea

As Huntington's chorea is inherited according to the laws of autosomal dominant inheritance, it could be expected that the affected parent of any offspring would, with equal probability, be the mother or the father. This is true with regard to adult-onset patients. However, it has been repeatedly shown that children affected with Huntington's chorea have an affected father approximately four times as frequently as an affected mother. The sex distribution of juvenile patients is, however, equal (Merritt et al. 1969; Myrianthopoulos 1973).

7.5.1. Predominance of Paternal Descent

The sex of the transmitting parent of persons with juvenile Huntington's chorea in different series is shown in Table 7.4. Went's results (1979) include those originally reported by Myrianthopoulos (1973) and are drawn from notifications from different parts of the world. Whilst the figures do vary, it is repeatedly evident that the father is more likely to transmit the gene to an affected juvenile than the mother. Furthermore,

Table 7.4. Sex of affected parent of patients with juvenile Huntington's chorea

Author(s)	No. of patients	Affected parent		Ratio	
		Father	Mother		
Merritt et al. (1969)	106	84	22	3.8 : 1	
Barbeau (1970)	33	26	7	3.7 : 1	
Went (1979)	94	78	16	4.8 : 1	
Hayden and Beighton (1981)	17	13	4	3.2 : 1	
	250	201	49	4.1 : 1	(av)

Table 7.5. Sex of affected parent of patients with childhood onset (0-9)

Author(s)	No. of patients	Affected parent		Ratio	
		Male	Female		
Myrianthopoulos (1973)	34	27	7	3.9 : 1	
Went (1979)	49	44	5	8.8 : 1	
Hayden and Beighton (1981)	6	5	1	5 : 1	
	89	76	13	5.8 : 1	(av)

Table 7.6. Sex of affected parent of patients with adolescent onset (10-19)

Author(s)	No. of patients	Affected parent		Ratio
		Male	Female	
Went (1979)	45	34	11	3.1 : 1
Hayden and Beighton (1981)	11	8	3	2.6 : 1
	56	42	14	3.0 : 1

division of the juvenile group into those with childhood onset and those who first show signs in the second decade reveals that the younger the age at onset the more likely it is that the transmitting parent is male (Tables 7.5 and 7.6). Further evidence is, however, needed to substantiate this latter finding.

Nevertheless, the constancy of this unexplained and unexpected finding with relation to the juvenile group as a whole suggests that it is unlikely to represent a bias in the ascertainment of data. In most instances affected parents were identified in large surveys in which no possible sex bias could have been exerted.

Numerous theories have been invoked which may account for this finding. Penrose (1948) has suggested that an allelic gene from the unaffected parent may precipitate an early onset of the disease in juvenile patients. However, the findings of Byers and Dodge (1967) and Delmas-Marsalet et al. (1968), who each described two juvenile patients who were half-brothers, being children of the same father and different mothers, renders this theory most unlikely. Myrianthopoulos (1973) has suggested that if modifying genes of the affected parent are responsible, they must somehow be sex-associated. It is obvious that these modifiers could not easily be assigned to either the X or Y chromosome as the sex distribution among juvenile patients is essentially equal. For this reason, these hypothetical modifiers are more likely to be located on one of the autosomes, although in some way not yet clearly understood they must also be sex-associated.

If the modifying gene were closely linked to the gene for Huntington's chorea, one would expect all sibs in a given kindred to inherit the linked modifer and thus also have early onset of symptoms. There are examples in given sibships where a child has developed signs of the disorder and his sibs have only shown features of Huntington's chorea in adulthood. On this basis, it is likely that the possible genetic factors precipitating early onset are not closely linked to the gene for Huntington's chorea.

Another possible explanation for the predominance of paternal descent in juvenile Huntington's chorea is that affected males who produce children with onset of the disease before the age of 20 are more fertile than similarly affected females. Merritt et al. (1969) first suggested that this mechanism might be operative, but this finding was later negated by Jones (1973) who showed that the women with Huntington's chorea are more fertile than the men at all ages of onset.

The consistent reports of the high frequency of paternal descent in the juvenile form of Huntington's chorea clearly reflect a genuine situation, which is not in accordance with classic Mendelian genetics and which remains a statistical fact with no adequate biological explanation.

7.5.2. Familial Aggregation

In view of the relative rarity of this condition, it is surprising to find that juvenile patients tend to aggregate in a particular kindred. Four affected sibs with juvenile Huntington's

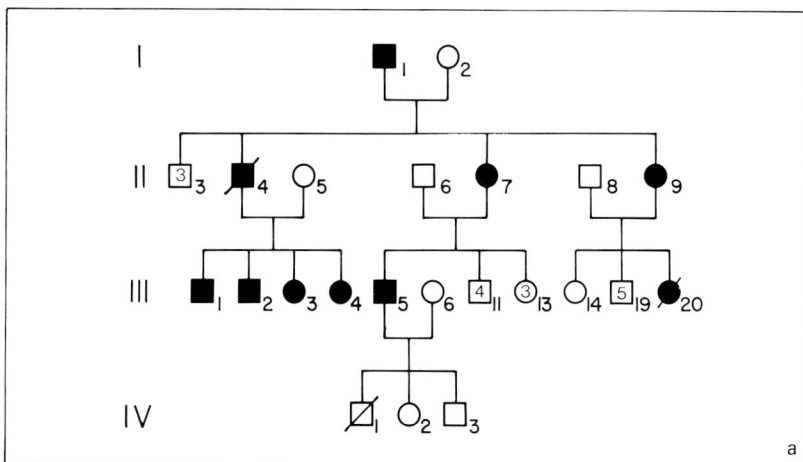

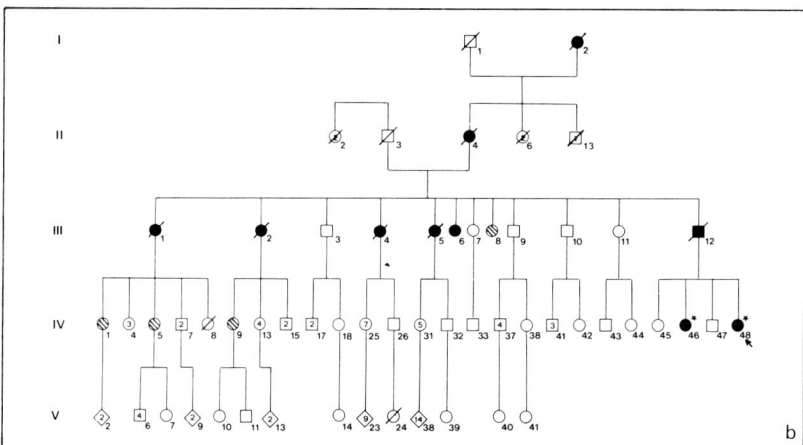

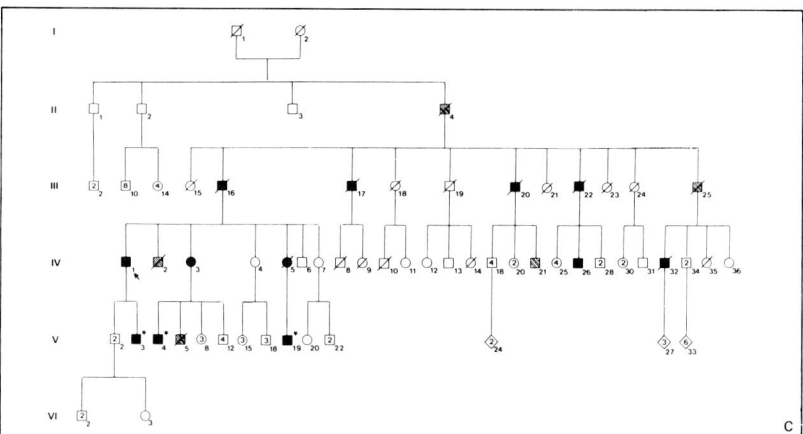

Fig.7.4. a Four sibs in generation III (III 1–4) have Huntington's chorea. **b** In generation IV two affected sisters, 46 and 48, both had onset before the age of 20. **c** The three affected juveniles in generation V are first cousins. (From Hayden and Beighton, 1981)

chorea are shown in the pedigre in Fig. 7.4a, whilst another two affected sibs (generation IV: 46, 48) and three affected first cousins (generation V: 3, 4, 19) are shown in Fig. 7.4b and c respectively.

Even though related sibs with juvenile Huntington's chorea were noted by Schmidt as early as 1892, and subsequently documented in many publications, the phenomenon of familial aggregation of juvenile Huntington's chorea has not been adequately discussed. It could be argued that environmental factors are responsible for this situation but they are unlikely to have a major influence as there are examples of sibs separated at birth into different environments who developed early onset of the disease. Whilst it would seem that genetic factors are most important in determining age of manifestation of the disease, this remains unproven.

7.6. Genetic Registers

A working party of the Clinical Genetics Society in Britain has recently proposed that genetic registers should be used in an effort to reduce the transmission of inherited disease (Emery et al. 1978). There are some who have advocated inclusion of Huntington's chorea in such a national register, as this may aid in early diagnosis and could serve as an integral part of the counselling services to affected families. However, this prospect has also caused concern in the minds of other persons aware of the problem of maintaining confidentiality and fearing possible authoritarian intervention in this area. If improperly used, such information could harm rather than improve the situation of the involved families. For example, unauthorised access could damage affected persons' chances of job promotion.

At the present time, there is a national register for Huntington's chorea in Denmark and America and a similar register is being planned for Australia (Chiu and Teltscher 1978), whilst lists of patients and their families are kept by centres interested in this disease in other parts of the world. In most areas, both affected persons and those at risk have been included in these records. From a logistical viewpoint such registers are easy to maintain and administer in countries with small populations such as Denmark but pose immense bureaucratic problems for larger nations where full-time workers would be needed to keep track of births, deaths and migration patterns.

Registers would be particularly useful when therapeutic advances and preventive measures become available as they would facilitate easy transmission of this information to all concerned. For this and other reasons mentioned previously, registers for Huntington's chorea should be maintained, but are best administered by specialised units with a particular interest in this disorder. Whilst benefits would accrue as a result of the establishment of a national register, this would only be acceptable once all fears concerning the possible breach of confidentiality and incursion into the patient's rights to privacy in a government-administered, national computer register had been allayed.

7.7. Genetic Counselling

In the past, the main thrust of genetic counselling was directed towards reduction of the frequency of inherited disorders. Genetic counselling is thus an offshoot of the eugenics movement, one of whose aims was to reduce the number of 'weak' or 'inferior' genes in the community. Eugenic measures for this disease were strongly advocated as far back as 1916, when Davenport and Muncey stated that it would be a work of far-seeing philanthropy to sterilize all those in which chronic chorea has already developed and to secure that such of their offspring as show prematurely its symptoms shall not reproduce. This is the least the state can do to fulfil its duty towards the as yet unborn. A state that knows who are its choreics and knows that half their children will become choreic and does not do the obvious thing to prevent the spread of this dire inheritable disease is impotent, stupid and blind and invites disaster.

Although latter-day geneticists do not identify themselves as eugenists, all would agree that a positive outcome of genetic counselling would be a decrease in the frequency of Huntington's chorea in the community. There have been no systematic studies of the effect of counselling measures on procreative behaviour in this disorder. Carter and Evans (1979) investigated 25 at-risk persons of reproductive age who had been counselled and found that the mean number of children for this group was 1.12, a little more than half-replacement rate. This may be early evidence of family limitation in response to counselling. Harper et al. (1979) are engaged in a long-term programme of non-directive genetic counselling in South Wales to all persons over 18 years of age at risk for Huntington's chorea, in an effort to determine whether such preventive measures have any significant effect on the future frequency of the disorder in their population.

The ideal of future generations free from Huntington's chorea will clearly never be realised, owing to the occurrence of spontaneous mutations and the impossible task of counselling all persons at risk and ensuring that they curtail reproduction. Nevertheless, this does not diminish the importance of attempts to decrease the frequency of the disease. Whilst efforts should still be directed to this end, it is evident that a broader, more holistic view of the function of the genetic counsellor in this disorder is necessary. The issue of whether the genetic counsellor represents the interests of the individual or of society at large is complex and a full discussion of this important problem is beyond the scope of this book. Briefly, it would seem, however, that the major emphasis of the past on the frequency of defective genes in the community has shifted to include the specific needs and concerns of the individual. There are some who argue that the latter's interest should receive priority (Milunsky 1975).

A definition of genetic counselling adopted by the American Work-group on Guidance and Counselling of the Commission to Control Huntington's Disease (Workgroup on Guidance and Counselling 1977) is as follows:

Genetic counselling is a communication process that has as its goal the alleviation of a genetic disorder in a family. The counsellor tries to achieve this goal by helping the counsellee:

1) Comprehend the medical facts, including the diagnosis, probable course of the disorder and the available management.

2) Appreciate the risk of recurrence in specified relatives.

3) Understand the options for dealing with the risk of recurrence.

4) Choose the option which seems appropriate to them in view of their risk and their values and act in accordance with that decision.

5) Make an optimal adjustment to the disorder, be that person affected or at risk of developing the disease.

In view of the relative uncommonness of Huntington's chorea the clinician may have had little experience of managing affected families. For this reason a paradigm for the provision of genetic counselling services is shown in Table 7.7. This scheme highlights some of the more important issues which should be explored during the counselling sessions. The content of counselling needs to be tailored to the particular concerns of the counsellee and, in some instances, it may be important to discuss issues planned for later interviews at an earlier stage. In many centres a multidisciplinary team of professionals is generally best equipped to manage these problems. Members of the group may include a neurologist, a psychiatrist, a geneticist and a social worker.

The specific problems of the patient, his spouse and at-risk relatives and an approach to their management will be described in Chaps. 8 and 9. It is clear that persons belonging to

Table 7.7. Paradigm for the provision of genetic counselling services

First interview

Make contact with family and establish rapport

Take a full family history

Examine the patient and make a preliminary clinical diagnosis
    Refer to appropriate sources if necessary

Explore patient's knowledge of the disorder and his/her psychological resources

Supply basic information concerning the disease and its consequences

Mention extensive research activity and encourage realistic hope

Second interview (1–2 weeks later)

Review information given and discuss problems and feelings arising from first interview

Supply further information about all aspects of the disease, with particular reference to caring facilities

Explore person's fears and fantasies in relation to Huntington's chorea

Reassess the diagnosis and evaluate the need for special investigations and therapy

Plan ongoing management, frequency of further meetings and referral to appropriate sources, dependent on
    patient's needs and concerns

Provide appropriate literature on disease

Issues for later interviews (interview interval depending on patient's needs)

Further information concerning the disease

Prospects for procreation and parenthood/reproductive options

Anticipation of possible future problems with affected person: depression, irritability, lowering of tolerance
    threshold

Legal aspects: wills, employment, insurance problems

Help the patient come to terms with his illness and formulate a future programme to deal with anxiety and
    fears of those at risk and relieve guilt of unaffected spouse

Discussion of varieties of care, rehabilitative services and their limitations

Preparation for death of affected person

Importance of autopsies for all affected persons

Help families to come out of hiding with this illness and accept it as any other long-term illness

Keep channels of communication open by stressing availability of self for periodic crisis support

each of these groups will need counselling which is based on their individual needs and concerns. The context of counselling and its mode of presentation will depend on their level of awareness and readiness to receive further details. Appropriate counselling can exert a considerable influence on the behaviour and well-being of the patient and his family. In most instances counsellees seek in counselling more than factual information (Kessler 1979). They are searching for some meaning in their experience with Huntington's chorea and through the counselling process may be encouraged to explore and discover this for themselves.

   Since one of the major aims of genetic counselling in Huntington's chorea is to ensure that decisions concerning reproductive behaviour are made in cognisance of all the relevant facts, a brief discussion of reproductive fitness in this disease is warranted.

7.8. Reproductive Fitness

Given the knowledge of the 50 per cent probability figure and the dreadful progression of the disease, the onset in middle life and the lack of a reliable predictor, there is no other disorder with such a strong argument against

Table 7.8. Fertility and reproductive fitness

Author(s)	Sex	Average number of children		Reproductive fitness
		Persons with chorea (A)	Normal sib (B)	
Palm (1953)	Total	5.1	2.7	1.9
Kishimoto et al. (1957)	Male	3.5		0.5
	Female	3.9		
	Total	3.7	5.7	0.6
Reed and Neel (1959)	Male	1.8	2.1	0.9
	Female	2.8	2.0	1.4
	Total	2.4	2.1	1.1
Wendt (1963)	Male	2.6	2.6	1.0
	Female	2.5	2.5	1.0
	Total	2.5	2.6	1.0
Wallace and Parker (1973)	Male	2.6	2.3	1.1
	Female	3.9	2.6	1.5
	Total	3.2	2.5	1.3
Marx (1973)	Male	2.4	2.5	1.0
	Female	3.6	2.6	1.4
	Total	3.0	2.5	1.2
Mattsson (1974)	Male	1.6	1.9	0.9
	Female	1.9	1.8	1.0
	Total	1.8	1.8	1.0
Stevens (1976)	Male	1.9	1.8	1.0
	Female	2.8	1.9	1.5
	Total	2.4	1.8	1.4

reproduction. Yet the potential victims do reproduce, apparently at a dysgenic rate. It is as if human reason and human emotion move on separate tracks, rarely accessible to each other. Perhaps the human drive for procreation springs from a well so deep that it will always override and take control of human intellect, blending all reason and argument into rationalisation through the process of rejection and denial (Pearson 1973).

The number of abnormal genes for Huntington's chorea transmitted from one generation to the next is in part a function of the fertility of the carriers of that gene. An unexpected result of most investigations of reproductive fitness in Huntington's chorea has been that affected persons generally produce more offspring than their normal sibs (Table 7.8; Fig. 7.5). Whether this enhanced fertility is a direct effect of the abnormal gene which confers enhanced biological fitness compared with its normal allele, as postulated by Wallace and Parker (1973), or rather the consequence of voluntary curtailment of reproduction on the part of those normal siblings is uncertain.

The notion that the abnormal gene has a biological selective advantage in the context of propagation compared with its normal allele, would be contrary to accepted notions concerning evolution. The determinants of fertility in Huntington's chorea are more likely to be behavioural than genetic. Decreased sexual inhibition coupled with lack of

MEAN FAMILY SIZE

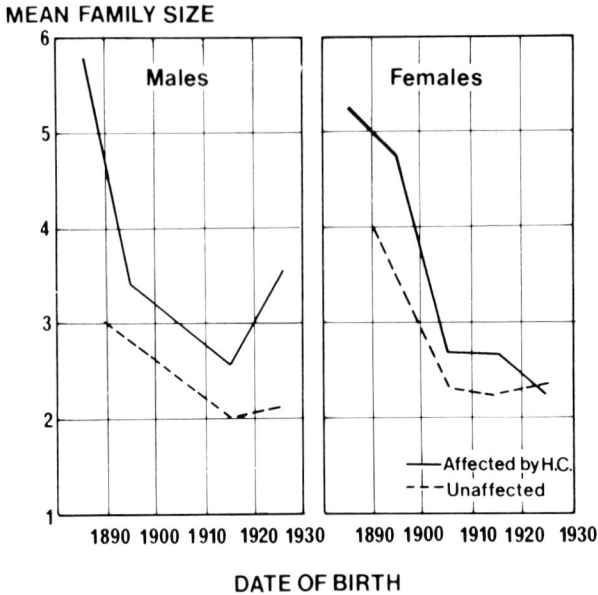

Fig.7.5. Mean family size in patients with chorea and their unaffected siblings. (From Harper et al. 1979)

competence in the use of contraceptive techniques on the part of the patient, associated with a voluntary limitation of procreation on the part of those at risk, could account for these findings.

When the general population is used for comparison, it is evident that males with Huntington's chorea are less fertile than their normal counterparts, affected females are more fertile, and the mean reproductive fitness of persons with this disorder, irrespective of sex, generally approximates that of the community. In other words, without preventive measures and notwithstanding mutations, the frequency of the disease will remain constant. Notable exceptions to this concept are the results of the studies of Marx (1973) and Stevens (1976), who found that the gene for Huntington's chorea enhances patients' fertility relative to the general population and that consequently, in the absence of adequate control measures, this disorder will become more common. The normal or increased reproductive fitness reported in Huntington's chorea is highly unusual for a severe, debilitating disease and the relative fitness rates are amongst the highest for any dominantly inherited disease (Reed and Neel 1959). The fact that the abnormal gene may confer a selective advantage defies biological explanation.

In contrast to the varying results relating to the comparison of fitness between persons with chorea and the general population, a remarkably consistent finding is that females with chorea are more fertile than similarly affected males (Tables 7.8 and 7.9). This is also a surprising finding for an autosomal disease of normal sex distribution. Whether this result is a true biological effect of the gene or reflects sociological variables, such as a decreased number of affected males who marry, with subsequently less procreation, is uncertain.

Another interesting finding is that the reported figures for reproductive fitness of Japanese affected persons are the lowest in the world. In a disease with a fixed mutation rate, this could be one of the factors responsible for the low frequency of Huntington's chorea in Japan. Comparison of the results of different studies of reproductive fitness is

Table 7.9. Reproductive fitness of patients with chorea compared with the general population

Author(s)	Reproductive fitness		
	Male	Female	Total
Reed and Neel (1959)	0.7	1.0	0.9
Wallace and Parker (1973)	0.9	1.1	1.0
Marx (1973)	0.9	1.4	1.2
Stevens (1976)	1.1	1.6	1.4

Adapted from Stevens (1976).

fraught with difficulty as varying methodological criteria for inclusion of persons have been used. Nevertheless, the preliminary findings in this area are sufficiently interesting to warrant further examination.

References

Avila-Giron R (1973) Medical and social aspects of Huntington's chorea in the state of Zulia, Venezuela. In: Barbeau A, Chase TN, Paulson GW (eds) Huntington's chorea 1872-1972. Raven Press, New York, pp 259-266

Barbeau A (1970) Parental ascent in the juvenile form of Huntington's Chorea. Lancet II: 937

Bell J (1934) Huntington's chorea. In: Fisher RA (ed) The treasury of human inheritance, vol IV/1. Cambridge University Press, London, pp 1-77

Byers RK, Dodge JA (1967) Huntington's chorea in children: a report of four cases. Neurology (Minneap) 23: 561-569

Carter CO, Evans K (1979) Counselling and Huntington's chorea (letter). Lancet II: 470-471

Chiu E, Brackenridge CJ (1976) A probable case of mutation in Huntington's disease. J Med Genet 13(1): 75-79

Chiu E, Teltscher B (1978) Huntington's disease: the establishment of a national register. Med J Aust 2: 394-396

Delhanty JH (1979) Communication to Eighth International Workshop on Huntington's Chorea, Oxford, England

Delmas-Marsalet P, Bourgeis M, Vital C, Fontanges T (1968) Formes rigides de la maladie de Huntington. Rev Neurol 118: 273-283

Eldridge R, O'Meara K, Chase TN, Donnelly EF (1973) Offspring of consanguineous parents with Huntington's chorea. In: Barbeau A, Chase TN, Paulson GW (eds) Huntington's chorea, 1872-1972. Raven Press, New York, pp 211-222

Emery AEH, Brough C, Crawford M, Harper P, Harris R, Oakshott G (1978) A report on genetic registers. J Med Genet 15: 435-442

Frederickson DS, Goldstein JL, Brown MS (1978) The familial hyperlipoproteinemias. In: Stanbury JB, Wyngaarden JB, Frederickson DS (eds) The metabolic basis of inherited disease, 4th ed. McGraw-Hill, New York, pp 604-655

Geratovitsch M (1927) Über Erblichkeitsuntersuchungen bei der Huntington'schen Krankheit. Arch Psychiatr Nervenkr 80: 513-535

Harper PS, Walker DA, Tyler A, Newcombe RG, Davies K (1979) Huntington's chorea: The basis for long-term prevention. Lancet II: 346-349

Hayden MR (1979) Huntington's chorea in South Africa. PhD thesis, University of Cape Town

Hayden MR, Beighton PH (1981) Genetic aspects of Huntington's chorea. Am J Med Genet (in press)

Heilbronner E (1903) Über eine Art progressiver Hereditat bei Huntington'schen Chorea. Arch Psychiatr Nervenkr 36: 889-894

Hindringer P (1936) Eine neue Chorea-Huntington-Sippe mit einer kurzen Zusammenstellung des gesamten Schrifttums der letzen 15 Jahre über Chorea Huntington. ID, Erlangen

Huntington G (1872) On chorea. Med Surg Rep 26: 317-321

Jelliffe SA (1908) A contribution to the history of Huntington's chorea: a preliminary report. Neurographs 1: 116-257

Jones MB (1973) Fertility and age of onset in Huntington's chorea. In: Barbeau A, Chase TN, Paulson GW (eds) Huntington's chorea, 1872-1972. Raven Press, New York, pp 171-177

Kessler S (1979) The psychological foundations of genetic counselling. In: Kessler S (ed) Genetic counselling: psychological dimensions. Academic Press, New York London, pp 17-35

Kishimoto K, Nakamura M, Sotokawa Y (1957) Population genetics study of Huntington's chorea in Japan. Annu Rep Res Inst Environ Med 9: 195-211

Kronthal P, Kalischer S (1895) Weiterer Beitrag zur Lehre von der pathologischanatomischen Grundlage der chronischen Chorea (hereditaria). Virchows Arch [Pathol Anat] 139: 303-318

Marx RN (1973) Huntington's chorea in Minnesota. In: Barbeau A, Chase TN, Paulson GW (eds) Huntington's chorea 1872-1972. Raven Press, New York, pp 237-249

Mattsson B (1974) Clinical, genetic and pharmacological studies in Huntington's chorea. UMEA University Medical Dissertations 7. UMEA, Sweden, pp 21-51

Merritt AD, Conneally PM, Rahman NF, Drew AC (1969) Juvenile Huntington's chorea. In: Barbeau A, Brunette JR (eds) Progress in neurogenetics, vol I. Excerpta Medica, Amsterdam, pp 645-650

Milunsky A (1975) Genetic counselling, principles and practice. In: Milunsky A (ed) The prevention of genetic disease and mental retardation. WB Saunders, Philadelphia, pp 64-87

Myrianthopoulos N (1973) Huntington's chorea: the genetic problem 5 years later. In: Barbeau A, Chase TN, Paulson GW (eds) Huntington's chorea, 1872-1972. Raven Press, New York, p 150

Osler W (1893) Remarks on the varieties of chronic chorea and a report upon two families of the hereditary form, with one autopsy. J Nerv Ment Dis 18: 97-112

Palm JD (1953) Detection of the gene for Huntington's chorea. PhD thesis, University of Minnesota, Minneapolis

Pearson JS (1973) Behavioural aspects of Huntington's chorea. In: Barbeau A, Chase TN, Paulson GW (eds) Huntington's chorea, 1872-1972. Raven Press, New York, pp 201-213

Pearson JS, Petersen MC, Lazarte JA, Blodgett HE, Kley IB (1955) An educational approach to the social problem of Huntington's chorea. Proc Mayo Clin 30: 349-357

Penrose LS (1948) The problem of anticipation in pedigrees of dystrophia myotonica. Ann Eugen (Lond) 14: 125-132

Reed TE, Neel JV (1959) Huntington's chorea in Michigan. II. Selection and mutation. Am J Hum Genet 11: 107-136

Reed TE, Chandler JH, Hughes EM, Davidson RT (1958) Huntington's chorea in Michigan: demography and genetics. Am J Hum Genet 10: 201-225

Schmidt A (1892) Zwei Fälle von Chorea Chronica Progressiva. Dtsch med Wochenschr 18: 585-587

Schultze F (1898) Über Poly-, Para- und Monoclonien und ihre Beziehungen zur Chorea. Dtsch Z Nervenheilkd 13: 409-421

Seip T (1928) Ein Fall von Chorea Huntington und einige Bemerkungen zu dieser Krankheit. Acta Psychiatr Scand 3: 139-152

Shokeir MH (1975) Investigations on Huntington's disease in the Canadian prairies. II. Fecundity and fitness. Clin Genet 7: 349-353

Smith H (1898) History of a case of Huntington's chorea. Med Rec (NY) 54: 422-423

Snyder LF, Doan CA (1944) Is the homozygous form of multiple telangiectasia lethal? J Lab Clin Med 29: 1211-1216

Stevens DL (1973) The classification of variants of Huntington's chorea. In: Barbeau A, Chase TN, Paulson GW (eds) Huntington's chorea, 1872-1972. Raven Press, New York, pp 51-64

Stevens DL (1976) Huntington's chorea: a demographic, genetic and clinical study. MD thesis, University of London, pp 1-338

Stevens D, Parsonage MJ (1969) Mutation in Huntington's chorea. J Neurol Neurosurg Psychiatry 32: 140-143

Vogel F, Motulsky AG (1979) Human genetics: problems and approaches. Springer, Berlin Heidelberg New York

Wallace DC, Hall AC (1972) Evidence of genetic heterogeneity in Huntington's chorea. J Neurol Neurosurg Psychiatry 33: 789-800

Wallace DC, Parker N (1973) Huntington's chorea in Queensland: the most recent story. In: Barbeau A, Chase TN, Paulson GW (eds) Huntington's chorea, 1872-1972. Raven Press, New York, pp 223-236

Wendt GG (1963) Die Fruchtbarkeit von Kranken mit Huntingtonschen Chorea. Proceedings of the Second International Congress of Human Genetics, vol II.

Went LN, Vegter-Van der Vlis M, Volkers W, Collewijn H (1975) Huntington's chorea. In: Went LN, Vermey-Keers C, van der Linden AGJM (eds) Early diagnosis and prevention of inherited disease. Leiden University Press, Leiden, pp 13-25

Went LN (1979) Communication to Eighth International Workshop on Huntington's Chorea, Oxford, England

Workgroup on Guidance and Counselling (1977) In: Report of the Commission for the Control of Huntington's Disease. Technical Report. US Government Printing Office, Washington, DC, pp 345-377

8   Living with Huntington's Chorea:
The Social Perspective

The psychosocial implications of Huntington's chorea are extremely important and largely unexplored. With no readily available cure, the major task in the management of this disease is to improve the quality of care. The purpose of this chapter is to provide a greater understanding of the social consequences of Huntington's chorea in the hope that this may highlight the unmet needs of the patient and his family. By so doing, efforts can then be concentrated on the particularly neglected areas in the management of this disorder.

There is, of course, no typical experience of Huntington's chorea. The affected person's reaction to the disorder is closely related to the premorbid personality and to the family and social support system. This, in turn, may be markedly influenced by the prior generation's experience of the illness. In other words, it is not helpful to view the social consequences for any affected person in isolation. A comprehensive family approach is needed, which embraces all members of the kindred as persons potentially in need of care. In broad principle the focus of concern of different members of each family is, to a large extent, influenced by their biological relationship to the abnormal gene. For this reason the social repercussions of this disorder for the affected person, those at risk and the unaffected spouse will be described separately. Phrases in inverted commas are extracts from conversations with affected persons and their families.

8.1.   Psychosocial Consequences for the Affected Person

The tragedy of this disorder is that it strikes in the prime of life when social responsibilities and personal and financial possibilities are greatest. Many patients in the earliest phase of their illness express great apprehension at being afflicted with "that dreadful hereditary disease, the slow killer which slowly and inexorably disintegrates the mind and body". As the disease progresses, fears ranging from realistic dangers of falling or self-injury to primal anxieties over "losing one's mind" may be consciously or unconsciously expressed. The inability to work, with subsequent economic impoverishment, compounds the devastating effects of this "disease of inherited madness".

Disturbances in speech despite adequate mental function, lead to difficulties in communication and frustration. It is particularly those persons with minimal intellectual dysfunction who may be acutely aware of their growing incompetence as a parent, husband or wife or sexual partner. It is frequently wrongly assumed that persons with disturbance in articulation have similar limitations in understanding. As a result, feelings of isolation may unwittingly be exacerbated by health professionals who explain details to, and answer questions from, other family members whilst communicating less with the patient. The inevitable result in the patient is depression and psychological withdrawal, with reluctance to express their own anxieties for fear of social abandonment.

In societies where a social stigma concerning mental disease prevails patients may be ostracised, not only by misinformed friends but also by their own family. One mother with fairly severe choreiform movements was prevented by her son from attending his wedding for fear of rejection by his friends and future parents-in-law.

Patients' despair is intensified by the knowledge that has accrued from watching "other members of their family wasting to an existence almost too horrible to contemplate". Loss of personal dignity may also be compounded by the failure of control of bladder and bowel function.

In spite of these problems, affected persons may be helped to attain a state of calm but realistic acceptance of the situation. Whilst this coping process is fashioned by the individual's beliefs and philosophy, a distinct pattern with predictable phases is often evident. An understanding of these stages facilitates appropriate counselling.

8.1.1. Psychological Defence Mechanisms

The presence of Huntington's chorea in the family invariably results in the development of different psychological defence mechanisms. Kübler-Ross (1970) has described the various psychological stages experienced by dying patients. In many respects these parallel the phases through which persons with Huntington's chorea pass in their unconscious attempt to achieve psychological homeostasis.

The earliest and most common defence employed by affected persons is denial, which may take many behavioural forms. The denial mechanism acts as a 'shock absorber' (Falek 1979a) which reduces the impact of the stressful situation and avoids the threat of reality. In some instances persons may state that "members of our family are all perfectly well" when they are fully aware of the family history of Huntington's chorea. Others may unconsciously distort information in an attempt to allay their own anxiety. "I have been told that blonde hair protects you from this disease. Therefore I have nothing to worry about." The negation of the existence of the disease is an attempt to safeguard employment and insurance possibilities, social and marriage prospects. However, denial may in a few situations be an honest representation of the facts, where parents may have died prior to onset of the disease or where the patient may be unaware of his own illegitimacy and the presence of Huntington's chorea in his natural parent.

Unfortunately persistence of denial may have serious implications as patients fail to come to terms with the disease and may actually procreate as a means of expressing their normality and healthy status. There are examples of persons with surnames associated with Huntington's chorea who have changed their name to that of a totally different culture, emigrated from their country of origin and given up all contact with relatives in an effort to "escape the family curse".

Whilst a realistic understanding of the situation for persons with Huntington's chorea is a desirable objective, it must be assessed at what expense to that individual's self and personal functioning this goal is to be achieved. Patients may see little value in affirming the presence of an illness for which they assume no therapy is available. An empathic attitude on the part of the counsellor which acknowledges the person's feelings is likely to help that individual move towards a realistic appraisal of the situation. Denial is usually superseded by phases of anxiety, anger and depression, and these behavioural changes may occur in any order. Whilst the expressed hostility may prove difficult to manage, it is the depression that may have the most serious consequences.

8.1.2. Suicide

'Insanity with a tendency to suicide' was one of the three features mentioned by Huntington (1872) in his description of the illness. It is often assumed that the high frequency of suicide in Huntington's chorea reflects the chronic, inexorably progressive nature of the illness (Wilson and Garron 1979). However, other chronic, incurable disorders such as multiple sclerosis have a much lower prevalence of suicide (Kurtzke et al. 1970), which suggests that other factors should be invoked to account for this finding.

Reed et al. (1958) have reported that 7.8% of deaths in non-institutionalised affected males and 6.4% of deaths in non-institutionalised females were due to suicide. Hayden et al. (1980) found that 3.35% of all deaths of affected individuals in South Africa were due to suicide. Various comparative estimates of the frequency of suicide in Huntington's chorea and the general population have been made (Falek 1979b; Hayden et al. 1980; Wexler 1977). The results vary depending on national suicide rates in different countries and show that suicide is between 7 and 200 times more common in Huntington's chorea than in the normal populaton.

Suicide is more frequent in the early phases of the illness when depression is marked and dementia is minimal. Affected males are more likely to commit suicide than affected females, whilst there is a greater frequency of attempted suicide in affected females. This parallels the pattern and sexual distribution of suicide in the general population. Those at risk for Huntington's chorea also commit suicide more commonly than the general population, although exact estimates for this group are not available. In many instances this arises as a result of misinformation concerning the disease.

It is evident that suicide may be considered a very real and valid strategy for some persons affected directly or indirectly by Huntington's chorea. A timely discussion of the patient's desires, fears and fantasies may be sufficient to allay their anxiety and avert this possibility.

8.1.3. Coping: Guidelines for Management

The broad principles for genetic counselling in Huntington's chorea have been described in Chap. 7. More specific guidelines which have been found useful when counselling affected persons will now be described.

Early, honest confrontation with the facts of the disease, with timely discussions concerning the problems that may ensue in the long-term, generates more constructive attitudes and maintains patient productivity commensurate with the stage of the illness. Many affected persons feel threatened by rejection and it is most reassuring for them to know that the physician is always available to answer any questions or provide treatment for problems that may arise. The fact that there are many persons in a similar situation in some way also lessens their feelings of isolation. Patients are encouraged by the knowledge that the research activity into all aspects of Huntington's chorea is greater now than it has been throughout the history of the disease. Even though no cure is available, there is treatment which can improve their symptoms and lessen their disability.

Regular follow-up in the absence of any urgent problems gives an opportunity for the patient to verbalise concerns and anxieties, and in this way caring for a patient with Huntington's chorea becomes a long-term commitment, with the clinician ensuring the continuity and availability of medical care.

With adequate and appropriate support patients learn to use their residual resources to their fullest potential and develop an attitude of acceptance. A daughter at risk explained how these channels were kept open right to the end with regard to her affected father.

"Till the last moment, even when he sometimes appeared in a coma, he still knew what was going on around him. My mother, who could still communicate with him, was aware of this and before he finally went to hospital, together they made plans for his own funeral." Another patient expressed his feelings in this way: "The most important thing is that everything I do from now on, I do with as much dignity as possible, so that my family remember me living with a little extra dignity."

8.2. The Experience of Being at Risk

Every child of an affected parent has an even chance of inheriting the gene for Huntington's chorea. The inability to escape from the unacceptable reality that they may be "passive victims of a totally random genetic accident" (Wexler 1979) is devastating for all concerned. A 28-year-old woman said "It's like living under a cliff waiting for a landslide." A man at risk, in a letter to his children, wrote as follows: "For all the joy and happiness I have had from you, my children, is clouded by the ever present dreadful 'if' always hanging overhead like a sword of Damocles. It clouds the reason and you tend to become conscious of every little action, wondering if this is the beginning of it." Others have described the experience of being at risk as "living with a time bomb" (Pines 1977), "playing Russian roulette with a two-barrelled gun and somebody else's hand on the trigger" (Wexler 1979).

The emotional response to being at risk takes different forms. Anger directed against the affected parent for transmitting this unwanted legacy is occasionally expressed. More often, however, feelings of warmth and sadness are voiced, as with the woman who said "I just felt overwhelming pity and love" when talking of her affected father. The possible transmisson of the gene for Huntington's chorea may be seen as retribution for some unsuspected wrong-doing. "I used to look at other families and wonder what I had done to deserve this. I only wanted to be normal."

Some persons feel that onset of the disease can be prevented or delayed by different methods, including special diet, prayer or yoga. In other instances individuals are less positive and become depressed as they are sure they will inherit the gene. Clumsiness, moodiness and a physical similarity to their affected parent are thought to be definitive pointers. Other family members may unconsciously reinforce these beliefs in the hope that they will be spared if other sibs are affected. We have on record a girl, aged 14, who was sterilised as a result of the family belief that she would develop Huntington's chorea. Her two siblings were not subjected to a similar procedure.

Some persons at risk continually check themselves in different ways for evidence of the disease. They may test their efficiency in simple tasks, ensure they can still walk along a straight line or assess their speed in performing jobs which require skills of coordination. In this way they become critical watchdogs of their own health.

Whilst the presence of Huntington's chorea may distort sibling relationships, it is the parent-child interaction that is most commonly fraught with conflict. Children are torn between their loyalties to each parent. On the one hand their sympathies are with their sick parent, whilst on the other they also understand the sacrificing, burdened spouse. Frequently the person at risk becomes the confidante of the well parent, who is often depressed and in need of relief. The children may urge the healthy parent not to renounce all their time and energy in the care of their sick spouse, whilst at the same moment realising that they are "encouraging the dreaded abandonment should they themselves become ill" (Wexler 1979). Their desire to provide optimal care for their parents is contrasted with their own needs for self-preservation. The unpleasant reality in these families is that there are too many persons in need of help with too few care providers.

Many individuals at risk may not routinely present for counselling as in their minds this is primarily the prerogative of the affected person. Yet it is particularly this group of people who are in need of such help as may enable them to reach responsible decisions concerning marriage, parenthood, employment, insurance and other major problems in their lives.

8.2.1. Some Guidelines for Counselling Those at Risk

Persons living with a parent with Huntington's chorea need preventive intervention which will facilitate coping with their own particular problems. Such persons often have no outlet for their own fears as the focus of caring is concentrated on the affected person. Regular interviews at prearranged time intervals, even when there is no specific crisis, provide a relaxed framework for the discussion and exploration of their problems.

When informing them of their at-risk status, a realistic assessment of the disease must be given. Nevertheless, the positive aspects of their particular situation are as worthy of emphasis as the negative ones. They have an even chance of not inheriting the gene, and the risk estimate can be modified by the methods described in Chap. 4. Furthermore, an encouraging aspect is that this disease has finally attracted the interest and attention of the medical profession.

If the clinician has examined the at-risk individual and is convinced that there are no signs of Huntington's chorea, it is reassuring for this person to be told that at the present time they show no features of the disease. It is useful to stress that the presence of clumsiness, depression, mood change, irritability and forgetfulness, which are often taken by those at risk as pointers to the presence of the disease, are normal concomitants of everyday life.

When a person learns of their at-risk status, this inevitably precipitates a crisis. A paradigm for the successful resolution of this crisis, which is applicable in other stressful circumstances, is shown in Figs. 8.1a and 8.1b.

Persons at risk for Huntington's chorea may be discriminated against in terms of employment and promotion possibilities. In some parts of the world, such as the USA, legislation has been enacted to protect them from such prejudice. The problem of eligibility for insurance is complex and unresolved. Those at risk are faced with the problem of whether or not to disclose their possible genetic inheritance, for fear of exclusion from insurance programmes. There are professionals who believe that the state should bear the financial burden of the increased premiums or settlements for affected families. Such a scheme, which would be applicable to similar genetic illnesses, would spread the cost of care to the entire population. At the present time, however, no such plan exists and life insurance is only obtainable for those at risk at substantially higher premiums than usual.

With appropriate intervention feelings of anxiety and fear can be displaced by an attitude of living more in the here and now. Such persons may have an intensified wish to fulfil their potential early in life, and this is amply evidenced by the large number of successful artists, businessmen, doctors, writers, musicians and others who, in spite of their at-risk status, are in the forefront of their particular field.

The most difficult and frightening decision for those at risk and the most frequent problem for the counsellor concerns marriage and procreation. In view of their major significance they will be considered separately below.

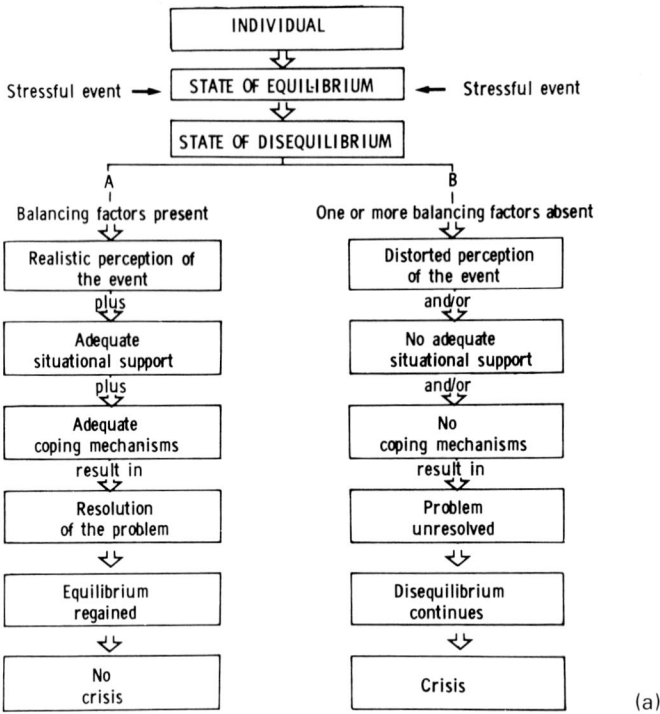

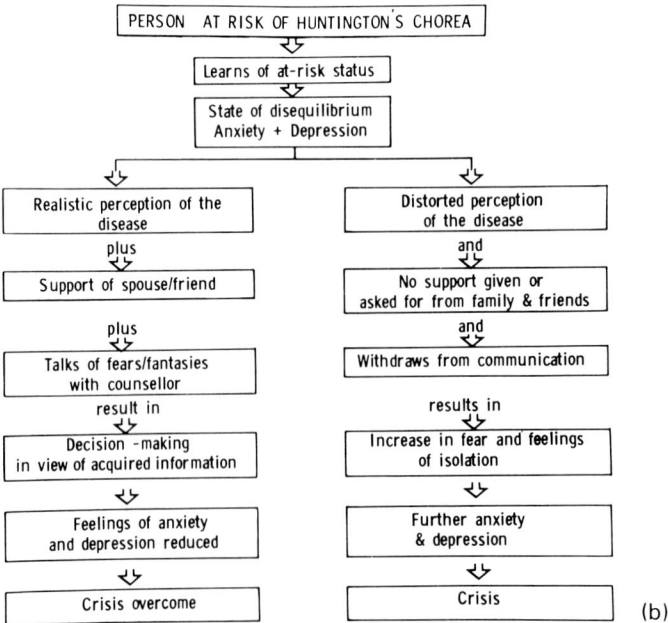

Fig.8.1.a A model of the balancing factors which alter one's response to a crisis situation. b The application of this model to persons at risk for Huntington's chorea. (Adapted from Aguilera and Messick 1978)

8.2.2. Marriage and Parenthood: The Options

The practising clinician can play a crucial role in the successful resolution of the dilemma concerning marriage and parenthood in Huntington's chorea. At-risk persons commonly present to the clinician for advice concerning procreation. In a few instances affected individuals may also request counselling in this regard. However, this is much less frequent in view of the delayed onset of the disease.

Whilst decisions concerning procreation for at-risk persons are intimately intertwined with those concerning marriage, it often helps to consider these issues separately.

A proportion of those at risk choose not to marry. For some the ambiguous situation of a 50% risk for this disorder has become a certainty that they will, in fact, develop the disease. They do not want to inflict the suffering and pain experienced by their unaffected parent on their own spouse. Their refusal to marry is thus closely related to their own perception and fantasies concerning Huntington's chorea and its consequences. Frequently, the denial of the option of marriage is closely related to the decision not to have children. In their minds, being at risk makes them undesirable as marriage partners and this is compounded by their refusal to procreate. The decision not to marry thus becomes a defence against future rejection.

Marriage is not to be discouraged, with the proviso that the spouse is informed of his/her future partner's at-risk status for Huntington's chorea. Whilst those at risk may fear this disclosure, the prospective spouse will usually appreciate this information, as decisions concerning procreation and other matters can then be made in full cognisance of the reality. Failure to acknowledge these facts may result in feelings of mistrust, resentment and alienation at a later stage. Sharing of this information lessens the isolation of those at risk, deepens the mutual understanding of the couple and allows them to plan together for responsible parenthood.

Approximately 25% of the at-risk individuals interviewed by Wexler (1979) procreated after learning of and comprehending the hereditary nature of the disease. Counsellors advising such couples on reproduction should be particularly mindful of their own personal views regarding childbearing in this disease. Whatever the counsellor's attitude, couples should be helped to find the decision that is right for them, which is based on their own needs and concerns and not the counsellor's. In the counselling process the potential parents can be urged, though, to explore their own reasons for procreation and to consider carefully the impact of their decision. However, once both persons decide to have children it is helpful to affirm this decision but to advise limitation of family size. If there is uncertainty, time should be spent and patience exercised in examining the different alternatives prior to decision-making.

If persons at risk choose not to have children, it is then the counsellor's task to help them to be reconciled to their childlessness and to explore other options for parenthood. The fact that one partner is at risk for Huntington's chorea considerably diminishes the couple's chances of becoming adoptive parents. In situations where there is a straight choice between a couple with neither partner at risk for Huntington's chorea and another with one partner at risk, the former couple usually takes preference. However, if the choice is between an institution or perhaps foster home and a couple at risk, it may be in the child's interest to be adopted by this couple. In view of the shortage of babies for adoption in many parts of the world, this latter situation is not a common occurrence.

Huntington's chorea poses other difficulties with regard to adoption. As a result of the social disintegration that may occur in some affected families, offspring may be given for adoption. In many instances the hereditary nature of the disease is not realised and the new parents adopt these children without an understanding of the implications of the disorder. The rights of prospective parents should, where possible, be protected and

Sterilization of children sought to stop disease

LONDON. — A British couple wants to have their three children sterilized because the mother suffers from a hereditary disease which they fear will be passed to future generations.

Mrs _____, 33, has Huntingdon's Chorea, a disorder of the nervous system which usually strikes in the 30s and 40s and causes jerking muscle spasms, slurred speech and premature ageing. There is no known treatment or cure for this and Mrs _____ said she had about five years to live.

"I know what a dramatic step this is, but because of their genes, my children have a 50-50 chance of contracting it," Mrs _____ told reporters at the family's home in the Staffordshire coal-mining town of Cannock.

"I saw my father die in pain, and my children are seeing me going toward my death in the same way, only by ending our line can we prevent them from watching their own children suffer."

The story of the parents' decision to seek sterilization for their children, _____, 14, _____, 11, and _____, 9, when they reach the age of 16, was given prominent play in London's newspapers. Sapa-AP

Fig.8.2. Article from a newspaper concerning sterilisation and Huntington's chorea

those persons adopting offspring of patients with Huntington's chorea should have a full understanding of the implications of the disease and be willing to accept the calculated risk.

A possible alternative to adoption arises if the potential carrier of the abnormal gene is a male. Artificial insemination of his wife would ensure that the gene is not transmitted to the next generation and at the same time allow her the fulfilment of pregnancy. However, this method raises numerous ethical and psychological issues which should be considered prior to its implementation. Another option which presents itself to at-risk males is to bank their sperm, undergo vasectomy and await the development of a predictive test or perhaps a cure. This alternative also poses many problems, primarily related to the uncertainty of advances in research in this area (Stern and Eldridge 1975).

There has been recent widespread publicity concerning persons at risk for Huntington's chorea who have been sterilised at a young age to prevent propagation of the gene (Fig. 8.2). Counsellors may be faced with the problem of young persons at risk who have decided on this course of action. In this situation it may be helpful initially not to focus on the issue of sterilisation, but to concentrate on that person's perception and experience of the disease. In some instances these persons may have a distorted view, having watched their parent die of Huntington's chorea, and be unaware of the improvements in research and care that are now available. In most countries termination of pregnancy may be offered to at-risk persons should contraceptive methods fail. These factors need serious consideration prior to making a decision in favour of sterilisation.

8.3.   The Burden on the Unaffected Spouse

The unaffected spouse has unique concerns and needs (Hans and Koeppen 1980). The change of marital role, with the husband having to deal with increasing domestic chores or the wife now being forced to become the breadwinner, is most difficult for the respective spouses. This problem is compounded by the inevitable isolation that ensues, following repeated social embarrassments and rejection by old friends. One woman, whose husband had been suffering from the illness for 5 years, described her anguish in this way:

> Nobody will help me. Is there nobody to give help when these things happen? Where can I go? I have to prepare food for my children. We have no money. What must I do? The psychiatrists said he has Huntington's chorea. He cannot sit still for a moment. He must be helped. Oh, this is a heart-breaking story. I have six children, three daughters and three sons. I heard this disease is inherited. What lies ahead of us? Who will be affected? I am exhausted. I cannot sleep.

In the past the majority of unaffected partners did not understand the implications of the hereditary nature of Huntington's chorea prior to marriage. The hardest fact for these persons to accept is that they have unwittingly been involved in the transmission of the disease to their children and many describe this guilt as their heaviest burden. The unaffected spouse is most often confronted with the invidious task of having to inform their children of the implications of their at-risk status. Often the cycle of misery continues after the death of their partner as they now witness the earliest signs and symptoms of the disease in some of their offspring.

The partner may often become the target of the patient's delusions and abuse and these escalating problems may seem so overwhelming that divorce becomes a realistic option. Whilst divorce ensues in some families, in many others the spouse chooses to stay with his or her partner. A feeling of acceptance and confidence is expressed. "I married him and it is my duty to look after him."

In the latter stages of the illness many patients are institutionalised, leaving the spouse alone. In some instances new relationships may be formed and the counsellor may be asked for advice regarding divorce. The unaffected spouse may not want to desert his or her partner, but may nevertheless feel guilt at their involvement with another person. Sympathetic listening and counselling help this person to reach their optimal decision.

Many persons complain of the indifference of the medical profession to their problems. "Doctors just seem to be not interested. The doctor at the hospital told me to take her [affected daughter] home as I probably know more about this Huntington's chorea than he does. What must I do? I go to the doctors for help and they can do nothing for me."

Whilst the afflicted patient may be the first focus of medical attention in these families, it is often the unaffected spouse who needs urgent help to cope with the emotional disruptions, economic burdens, curtailment of social activities and other endless crises. Counselling will help these persons contend with the escalating problems of the disease. Membership of lay organisations will also lessen their isolation and provide a forum for discussion of mutual concerns; this will be further discussed in Chap. 9.

8.4.   Huntington's Chorea: A Family Disease

The theme of this chapter has been that Huntington's chorea is a prime example of a family disease. This is poignantly described by a member of an affected family as follows. "To watch one's loved ones suffering this disease is a hell-on-earth existence. We, his family, are suffering as much as he is. The heartache it brings is worse than any physical

Fig.8.3. A photograph showing the patient surrounded by his wife, six children, five grandchildren and the dog

pain." The repercussions of the disorder are often felt beyond the boundaries of immediate blood relations. More distant relations, neighbours and friends are involved in the provision of care and in sharing the social and economic burden on the affected family.

A family photograph is shown in Fig. 8.3. The patient (middle of back row) is surrounded by his children and grandchildren, all of whom are affected by the social and biological effects of this disease. The problems are particularly acute in underprivileged sections of society, where there is no financial support and less access to community resources. Such a family is shown in Fig. 8.4. The father, who was the breadwinner until the onset of his illness, now needs daily care which prevents his wife from seeking employment. Lack of adequate social security benefits and medical insurance aggravates their plight.

8.5. Antisocial Behaviour: Huntington's Chorea and the Law

It is on record that the earliest transmitters of the abnormal gene to the USA had problems with the legal system of their adopted country as a result of repeated crimes and misdemeanours (Vessie 1932). Since that time there have been numerous reports of antisocial behaviour, including minor crimes such as offences against property and theft and more major offences such as assaults, murder and sexual aberrations (Bolt 1970; Hayden et al. 1980; Parker 1958). Dewhurst et al. (1970) noted that approximately 18% of their survey population were involved in legal proceedings on charges ranging from

Fig.8.4. Huntington's chorea has aggravated this family's plight by preventing the affected father from being employed. The mother stays at home to provide continuous care

cruelty to child abuse. A further 19% were dependent on alcohol. Oliver (1970) studied the clinically unaffected siblings of Huntington's chorea victims and found that of 150 persons, 17 died before the age of 11 while 9 died between 11 and 21 years. These figures may reflect a small number of undiagnosed patients, but in a large measure are evidence for the presence of ill-treatment and neglect in these families.

The presence of antisocial behaviour in Huntington's chorea is clearly established, but the precise determinants of such conduct are more difficult to ascertain. Whether these misdemeanours are the results of the primary underlying defect of the disorder or rather the consequence of the disturbed social environment is uncertain.

F. Baro (1979, personal communication) has stated that patients with Huntington's chorea commit major crimes more frequently than do the general population and that this is probably the consequence of disinhibition or lack of control of aggressive impulses, which is in part the result of the degenerative process of the disease. The finding that many patients who perform antisocial acts come from disturbed home environments suggests that the socially impoverished surroundings certainly contribute to the causation of these acts. Whilst this viewpoint is generally accepted, further studies in which the unaffected members of the family, other chronic neurological patients and members of the general population are used as controls are needed to compare the frequency of major and minor crimes in these different groups.

It is important that the legal profession and prison authorities be made aware of Huntington's chorea and its social implications. The presence of this disorder in an accused raises the possibility of diminished responsibility at the time of the crime. Caro and Haines (1973) stated that "we have been persecuting this particular group of unfortunates for a very long time, but now instead of burning them at the stake, we simply lock them up for rape or attempted murder, rather than understand that their problems are not of their own making". This situation is likely to improve in societies where programmes of education concerning Huntington's chorea are implemented.

8.6. The Economic Burden

Huntington's chorea imposes a significant financial burden on both the family and the community. The inevitable recurrence of this disease in successive generations imposes material hardships on affected families which create a cycle of economic disadvantages that often improverishes them.

Because of its clinical variation and different family settings it is impossible to calculate a figure that meaningfully reflects the costs of Huntington's chorea in every situation. Nevertheless, a minimum estimate can be computed for different countries which takes into account usual periods of hospitalisation, pharmacotherapy, other modes of treatment, such as counselling, and possible state aid. Notably, these figures generally exclude one of the largest single economic costs to the family, namely that of loss of earnings. Other expenses such as that of extra housekeeping, special safety devices, aids to the handicapped and other rehabilitative services are generally not included.

The Commission for the Control of Huntington's Disease and its Consequences in the USA (1977) computed figures of between $65 000 and $234 000 for the direct lifetime cost per patient in different scenarios, which illustrates the range of resources used by affected persons. A comparable minimum British estimate is £20 000 whilst the South African equivalent is approximately R28 000. The cost of Huntington's chorea in the USA per year ranges from $110 million to $125 million for direct costs as mentioned above. The social service caring costs to the National Health Service in Britain are estimated to be in the order of £4 million per annum (Office of Health Economics 1980). These totals were based on conservative assumptions and are minimum figures which can be regarded as understatements of the real cost in those countries at the present time.

Costs of this magnitude clearly show the potential savings to health services to be gained from a reduction in the frequency of this disorder. In the USA the only way some families have been able to subsist has been for the spouse to divorce the affected person, thus making him/her a dependant of the state. Calculation of the cost of this disease serves different important purposes. Harper et al. (1979) have estimated the cost/benefit ratio of their preventive programme and shown that it would be cost-effective if it reduced the birth rate of their population at risk by 10%. Individual agencies, government bodies and private foundations are often unaware of the economic problems confronting families with Huntington's chorea. These cost figures may be used to impress upon those in legislative power the financial devastation of Huntington's chorea in an effort to mobilise economic aid for affected families.

There has been a welcome shift in emphasis on the part of health professionals in dealing with the social consequences of this disease. In the past a passive attitude of waiting until problems occurred and then attempting to deal with them was adopted. However, as a result of the growing awareness and dissemination of information concerning Huntington's chorea, an active search for potential problems prior to their occurrence, followed by preventive intervention, is more commonly being undertaken. The counsellor whose emphasis is on the coping, integrative and adaptive functions of the personality has a particularly salient role in such preventive efforts (Cain 1977).

References

Aguilera DC, Messick JM (1978) Crisis intervention: theory and methodology, 3rd edn. CV Mosby, St. Louis

Bolt JM (1970) Huntington's chorea in the West of Scotland. Br J Psychiatry 116: 259-270

Cain A (1977) In: Report to the Commission for the Control of Huntington's Disease and its Consequences, vol II, Technical Report, US Government Printing Office, Washington, DC, pp 473-497

Caro A, Haines S (1973) Crime and psychiatry. Justice of the Peace: 621-622

Dewhurst K, Oliver JE, McKnight AC (1970) Socio-psychiatric consequences of Huntington's disease. Br J Psychiatry 116: 255-258

Falek A (1979a) Observations on patient and family coping with Huntington's disease. Omega 10(1): 35-42

Falek A (1979b) Communication to Eighth International Workshop on Huntington's Chorea, Oxford, England

Hans MB, Koeppen AH (1980) Huntington's chorea: its impact on the spouse. J Nerv Ment Dis 168: 209-214

Harper PS, Tyler A, Walker DA, Newcombe RG, Davies K (1979) Huntington's chorea: The basis for long-term prevention. Lancet II, 346-349

Hayden MR, Ehrlich R, Parker H, Ferera SJ (1980) Social perspectives in Huntington's chorea. S Afr Med J 58: 201-203

Huntington G (1872) On chorea. Med Surg Rep 26: 317-321

Kübler-Ross E (1970) On death and dying. Tavistock, London

Kurtzke J, Beebe G, Nagler B, Nefzger MD, Auth T, Kurland T (1970) Studies on the natural history of multiple sclerosis. V. Long-term survival in young men. Arch Neurol 22: 215-225

Office of Health Economics (1980) Huntington's chorea. OHE Paper 67. OHE, London

Oliver JE (1970) Huntington's chorea in Northamptonshire. Br J Psychiatry 116: 241-253

Parker N (1958) Observations on Huntington's chorea based on a Queensland survey. Med J Aust 45: 351-359

Pines M (1977) Living with a time bomb. In: Commission for the Control of Huntington's Disease and its Consequences, vol II, Technical report. US Government Printing Office, Washington, DC, pp 383-397

Reed TE, Chandler JH, Hughes E, Davidson RT (1958) Huntington's chorea in Michigan: demography and genetics. Am J Hum Genet 10: 201-225

Stern R, Eldridge R (1975) Attitudes of patients and their relatives to Huntington's chorea. J Med Genet 12: 217-223

Vessie PR (1932) On the transmission of Huntington's chorea for 300 years. The Bures family group. J Nerv Ment Dis 76: 553-573

Wexler NS (1977) In: Commission for the Control of Huntington's Disease and its Consequences, vol IV. Government Printing Office, Washington, DC, p 71

Wexler NS (1979) Genetic 'Russian roulette': the experience of being 'at risk' for Huntington's disease. In: Kessler S (ed) Genetic counselling: psychological dimensions. Academic Press, New York London, pp 190-220

Wilson RS, Garron DC (1979) Cognitive and affective aspects of Huntington's disease. In: Chase TN, Wexler NS, Barbeau A (eds) Advances in neurology, vol 23. Raven Press, New York, pp 93-203

9   Management

No treatment seems to be of any avail and indeed nowadays its end is so well known... that medical advice is seldom sought. It seems at least to be one of the incurables. (George Huntington 1872)

Huntington's chorea
Means there's no help known,
In the science of medicine
For me
And all of you, Choreanites, like me
Because all of my good nurses
And all of my medicine men
And all of my good attenders
All look at me and say
By your words and by your looks
Or maybe by your whispers
There's just not no hope
Nor not no treatments known
To cure me of my dizzy
Called Chorea.
  (Woody Guthrie, November 1954
  Brooklyn State Hospital, New
  York State. By courtesy of
  Marjorie Guthrie)

At the present time there is still no cure for Huntington's chorea, nor is there any therapy which significantly alters the natural progression of this disease. Nevertheless there are numerous aspects of this disorder which may be improved by appropriate medical intervention. In the absence of a cure, amelioration of the patient's symptoms becomes a desired and realistic objective. Although the precise pathogenetic mechanisms in Huntington's chorea are yet to be determined, recent advances in neuropharmacology have facilitated a more rational approach to the pharmacotherapy of the disease and this will be reviewed here. There is a wide range of other therapeutic modalities that may be beneficially used in the management of Huntington's chorea (Fig. 9.1) and a brief résumé of these services will also be presented.

9.1.   A Rational Approach to Pharmacotherapy

9.1.1.   Chorea

It has been suggested that hyperfunction of the nigro-striatal dopamine system accounts for the choreiform movements in Huntington's chorea (Chase 1976). In general drugs

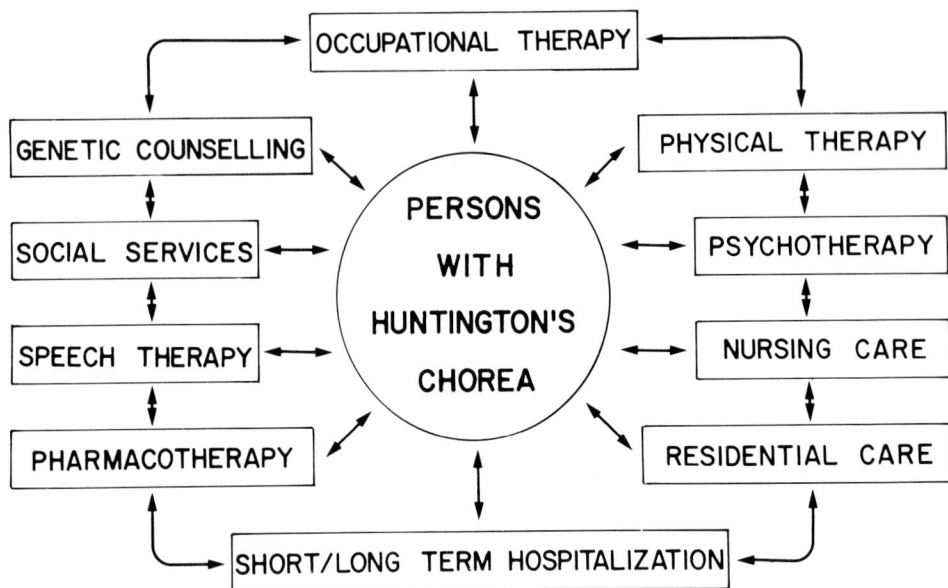

Fig.9.1. A whole range of therapeutic services may benefit patients with Huntington's chorea

which increase dopamine concentration in the central nervous system, such as L-dopa, worsen chorea (Klawans et al. 1972), whilst those pharmacological agents which block dopaminergic function (phenothiazines and butyrophenones) tend to suppress these involuntary movements.

The first problem confronting the practising clinician is to determine whether any therapy for chorea is necessary. If the movements are mild and do not limit normal function the possible side effects of such therapy may be more troublesome to the patient than the chorea itself. However, in most situations drug therapy will be indicated. The drugs that have been shown to consistently afford relief from chorea and their postulated mode of action are shown in Table 9.1. In general these pharmacological agents operate either by depleting the brain of monoamines or by blockade of dopamine receptors (Fig. 9.2). No specific drug has been shown to have a consistent superiority over others in the amelioration of chorea and the choice of first-line therapy will depend on the clinician's experience and the availability of drugs in that particular country. Certain broad guidelines for implementation of pharmacotherapy in Huntington's chorea are shown in Fig. 9.3.

The Committee to Combat Huntington's Disease in the USA recently released a schedule (compiled by Fahn 1979) regarding preferred medications in this disorder. Perphenazine was suggested as the drug of choice in view of its positive effect of reducing the abnormal movements. Its low price and infrequent side effects are additional advantages. Haloperidol, a butyrophenone, and reserpine, a central monoamine depleter, were suggested as possible alternatives. In the United Kingdom and Australia tetrabenzine, a synthetic benzoquinoline, has been used with beneficial effects (Dalby 1969; Kingston 1979; Taglia et al. 1978), and in the former country it is often the first drug of choice and the most widely used. Tetrabenzine is not available in the USA.

In general, more common side effects of these agents include drowsiness, depression and a Parkinsonian syndrome. Difficulty with concentration and feelings of obtundation

Table 9.1.   Drugs which reduce the involuntary movements of Huntington's chorea

Class	Drug	Rec. starting dose[a]	Mode of action	Author(s)
Phenothiazines	Perphenazine	4 mg tds	Postsynaptic dopamine receptor blockade	Fahn (1972, 1973)
	Fluphenazine	2 mg daily		Whittier and Korenyi (1968)
	Trifluoperazine	2 mg tds		Stokes (1975)
	Chlorpromazine	25 mg tds		Chase (1976)
Butyrophenones	Haloperidol	1.5 mg bd	Postsynaptic dopamine receptor blockade	Barbeau (1973)
Benzoquinolenes	Tetrabenzine	25 mg bd	Depletes central monoamine stores	Kingston (1979)
Alkaloids	Reserpine	0.25 mg daily	Depletes central monoamine stores	Lazarte et al. (1975); Ringel et al. (1973)

[a] bd, twice daily; tds, thrice daily

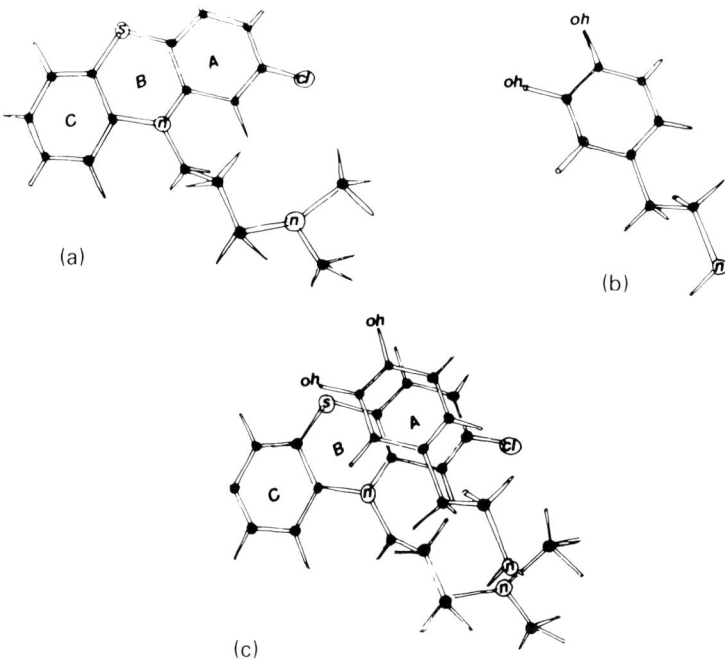

Fig.9.2. Drawings of models of the molecular structures of: A, chlorpromazine; B, dopamine; C, dopamine, superimposed on a portion of the chlorpromazine molecule. Chlorpromazine blocks dopamine receptors, which is understandable in view of their structural similarity. (Adapted from Horn and Snyder 1972)

```
┌──────────────────────────────┐   ┌──────┐   ┌──────────────────┐
│ DECIDE IF THERAPY IS NEEDED   │──▶│  NO  │──▶│ REASSESS LATER   │
└──────────────────────────────┘   └──────┘   └──────────────────┘
              │
              ▼
          ┌───────┐
          │  YES  │
          └───────┘
              │
              ▼
┌──────────────────────────────┐
│ CHOOSE DRUG WHICH IS KNOWN    │◀──────────────────────┐
│ TO PRODUCE A POSITIVE EFFECT  │                       │
└──────────────────────────────┘                       │
              │                                         │
              ▼                                         │
┌──────────────────────────────┐                       │
│ START WITH A LOW DOSE         │                       │
└──────────────────────────────┘                       │
              │                                         │
              ▼                                         │
┌──────────────────────────────┐   ┌──────────────────────────────┐
│ INCREASE DOSAGE CAUTIOUSLY    │   │ POOR / NO RESPONSE /          │
│ UNTIL MAXIMAL THERAPEUTIC     │──▶│ TROUBLESOME SIDE EFFECTS      │
│ AFFECT ACHIEVED, OR SIDE      │   └──────────────────────────────┘
│ EFFECTS INTERVENE             │
└──────────────────────────────┘   ┌──────────────────────────────┐
              │                     │ ASSESS PROGRESS OBJECTIVELY   │
              ▼                     │ BY TESTING FUNCTIONAL         │
┌──────────────────────────────┐   │ ABILITY IN SIMILAR TASKS      │
│ ASSESS PATIENT AT REGULAR     │◀──│ ON CONSECUTIVE OCCASIONS      │
│ INTERVALS . EVALUATE          │   └──────────────────────────────┘
│ PROGRESS, ADJUST DOSAGE       │
└──────────────────────────────┘
              │
              ▼
┌──────────────────────────────┐
│ IF PROGRESS EQUIVOCAL,        │
│ ADVISABLE TO DISCONTINUE      │
│ ALL THERAPY AND REASSESS      │
│ MOTOR STATUS                  │
└──────────────────────────────┘
```

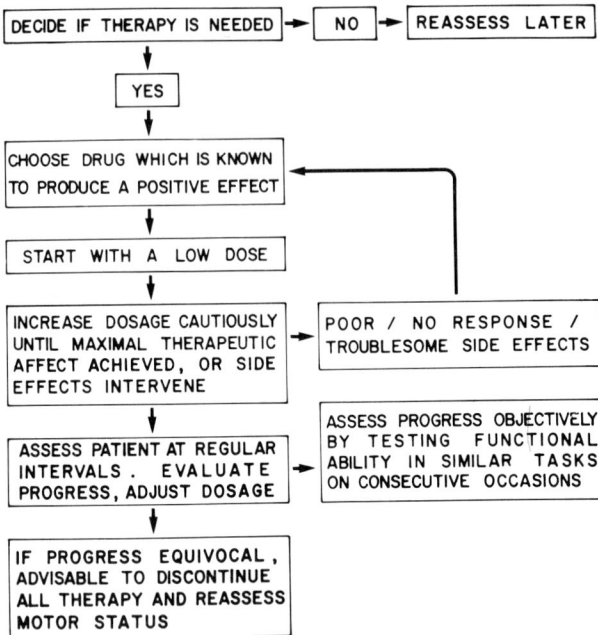

Fig.9.3. Guidelines for pharmacotherapy in Huntington's chorea

may significantly impair the patient's lifestyle and may be wrongly attributed to progression of the disease. Close attention to these problems, with dosage or drug change, may be indicated in these instances. Less common side effects may include dry mouth, constipation and urinary retention, and are due to the anticholinergic actions of these agents. Tardive dyskinesia has not been clearly documented as a side effect of dopamine antagonist therapy in Huntington's chorea. However, this is probably due to the inherent difficulty in establishing such a diagnosis rather than its non-occurrence. In summary, even though side effects are generally mild, great discretion is required in the use of these agents. In all instances dosage should be reduced to the lowest level compatible with symptom control.

Biochemical evidence of cholinergic neurone involvement is based on the finding of a marked reduction in choline acetyltransferase, the enzyme catalysing the conversion of choline to acetylcholine, in the brains of patients with Huntington's chorea (Table 10.5). However, therapy aimed at increasing cholinergic influence in the basal ganglia of affected persons has been disappointing. Administrations of choline (Aquilonius and Eckernas 1977; Growden et al. 1977) and arecoline produce elevations in blood and brain choline concentrations, but do not significantly improve the abnormal movements. Physostigmine and other cholinesterase inhibitors (Tarsy et al. 1974) may produce a transient improvement in patients' symptoms but this also is not sustained. The failure of cholinergic enhancement to improve the symptoms in Huntington's chorea probably reflects damaged cholinergic neurones which are thus not responsive to increases in brain choline levels.

The discovery of diminished γ-aminobutyric acid (GABA) concentration in the brains of patients with Huntington's chorea (Perry et al. 1973) led to therapeutic endeavours to augment brain GABA concentration. Different investigative approaches have been tried

Table 9.2. GABA-ergic therapy in Huntington's chorea

Mode of action	Drug	Effect	Author(s)
GABA-mimetic drugs			
	GABA	Nil	Barbeau (1973)
Receptor agonist	Muscimol	Nil	Shoulson et al. (1978)
Receptor agonist	Imidazole-4-acetic acid	Nil	Shoulson et al. (1975)
Receptor agonist	Baclofen	Nil	Anden et al. (1973)
Receptor agonist	SL-76002	Nil	Agid (1979)
GABA precursors			
	L-glutamate	Nil	Barr et al. (1978)
GABA transaminase inhibitors			
Also inhibits succinic semi-aldehyde dehydrogenase	Dipropylacetic acid	Nil	{ Bachman et al. (1977) Shoulson et al. (1976)
Also inhibits GAD	γ-acetylenic GABA	Nil	Agid (1979)
Also inhibits GAD	γ-vinyl GABA	Nil	Agid (1979)
Also inhibits GAD	Isoniazid	Improvement in some	Perry et al. (1979)
	Aminooxyacetate	Nil	Perry et al. (1980)

and these are summarised in Table 9.2. GABA-mimetic drugs and other pharmacological agents which interfere with GABA transaminase and succinic semialdehyde dehydrogenase enzymes facilitating the catabolism of GABA have been used (Fig. 9.4). In this way, it was hoped to increase brain GABA-ergic activity. With the possible exception of isoniazid, which in doses three to five times greater than those used to treat tuberculosis may reduce chorea, these drugs have not been helpful in relieving the symptoms of Huntington's chorea. Until subjected to controlled double-blind trials, the beneficial action of isoniazid in this situation also remains unproven.

The benzodiazepines have been found to be useful adjuncts to the therapeutic regimes of certain patients. It is thought that this positive effect is mediated through their sedative action and also possibly as a result of their facilitation of GABA-ergic transmission (Enna and Maggi 1979).

Attempts have also been made to treat Huntington's chorea with lithium. Positive results were reported in nine non-blind studies, whilst five well-controlled double-blind trials demonstrated no improvement with such therapy (Schou 1979). There is no rationale for the use of lithium treatment for the movement disorder of Huntington's chorea. Other drugs, including corticosteroids, have also been used with equivocal results (Brown et al. 1979).

Shortly after Corrodi and his colleagues (1973) discovered that bromocriptine had central dopamino-mimetic activity, Calne and associates (1974) reported that it had anti-Parkinsonian activity. It was therefore surprising that this same drug was shown to benefit patients with chorea. Recently Frattola and his co-workers (1977) demonstrated that low doses (<40 mg) of bromocriptine produce a marked improvement in Huntington's chorea. Other investigators, including Loeb et al. (1979) and Albano and Cocito (1979), have confirmed these initial findings, whilst Kartzinel et al. (1976) were

GLUCOSE

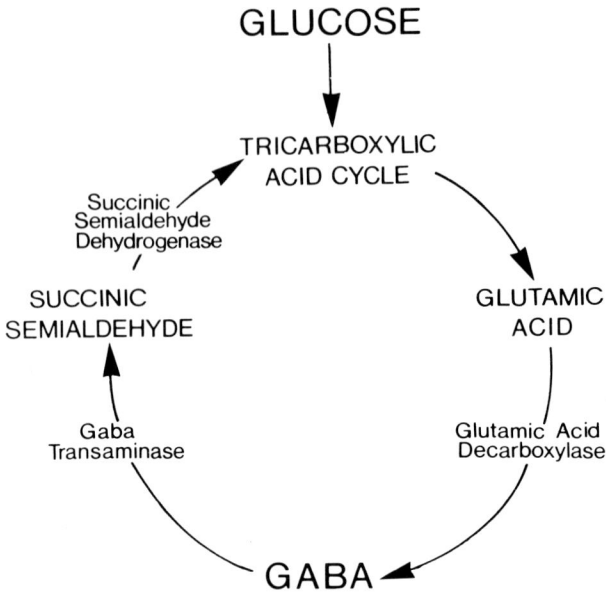

Fig.9.4. Metabolism of GABA

unable to show a similar beneficial effect. Amelioration in patients' movements has also been reported with low doses of another dopamine agonist, namely apomorphine (Corsini et al. 1978; Tolosa and Sparber 1974). These phenothiazine-like effects are presumably mediated by stimulation of presynaptic dopamine receptors, namely autoreceptors which result in decreased dopamine release from nerve terminals (Carlsson 1976). In higher doses both these dopamine agonists exacerbate chorea. These differential effects according to dosage are incompletely understood and at the present time these agents offer no therapeutic advantage over the accepted lines of treatment.

Recent advances in pharmacology have resulted in the introduction of a new class of dopamine antagonists, the substituted benzamides, which includes drugs such as sulpiride, tiopride and metoclopramide (Jenner and Marsden 1979). These drugs have more specific effects than the neuroleptics with less troublesome side effects and may prove to be of particular benefit in the therapy of Huntington's chorea (Zanglein et al. 1978).

It is perhaps discouraging that in spite of recent advances in knowledge concerning the neurochemistry of Huntington's chorea (Sect. 10.7), the most efficient drugs for the control of chorea are still the dopamine antagonists. In view of the complex feedback arrangements of striatal interneurones and the interaction of different neurotransmitters it is theoretically feasible that pharmacotherapy in Huntington's chorea could be more effective if it did not focus exclusively on one neurotransmitter system. Shoulson (1979) has proposed consideration of simultaneous or serial drug administration on this basis. Notwithstanding the fact that the crucial biochemical defect of Huntington's chorea still eludes us the ongoing research into the neuropeptides and other neurotransmitters is likely to lead to improvements in the pharmacotherapy of this disease.

9.1.2. Rigidity

The predominant motor abnormality in a small proportion of patients with Huntington's chorea, particularly those with juvenile onset, is rigidity. The drugs used for chorea usually exacerbate these patients' hypokinetic features. Persons with rigidity of an extrapyramidal nature may improve on conventional anti-Parkinsonian therapy, including L-dopa and the anticholinergic drugs. Careful monitoring of these patients' response to therapy should be maintained. At present it is difficult to develop a rationale that logically explains the simultaneous presence of symptoms due to both dopamine excess and dopamine deficiency in patients of the same family who have a common genetic defect. In this situation the plan for therapy is empirically derived.

9.1.3. Epilepsy

Epilepsy is uncommon in affected adults, but may occur with greater frequency in children with the disease. These seizures may be particularly resistant to conventional anti-epileptic therapy and high doses are usually needed to control this symptom. With these doses, close and regular monitoring for possible side effects is mandatory.

9.1.4. Dementia

The intellectual impairment of Huntington's chorea is not amenable to pharmacotherapy. However, drowsiness, with impaired mental function, may be a side effect of drug therapy and may show some improvement on cessation of treatment. However, in spite of the lack of suitable pharmacotherapy, simple measures can be taken which will often improve the affected person's well-being and extend their independence. Emphasis ideally should be placed on maintaining a familiar environment. A schedule of daily events, with fixed times for walking, eating and other activities such as visiting, can easily be constructed. Signs reminding the patient where he or she is, with a calender clearly displaying the year, helps prevent anxiety that may result from disorientation. These simple devices can influence and improve the patient's quality of life.

9.1.5. Affective Disturbance

Affective disturbance is common in Huntington's chorea and many depressed patients respond to conventional antidepressant medication with tricyclics (Chiu 1979). The principles of therapy with these drugs are similar to those outlined in Fig. 9.3. These are to begin with low dosage and gradually increase the dosage until a clinical change is observed; then to maintain adequate drug levels for at least 21 days prior to assessing the efficacy of such therapy. There have been no reports of carefully controlled trials of the use of antidepressant medication in this disorder and such investigations are clearly needed.

9.1.6. Psychosis

In general, the psychosis of this disorder responds favourably to conventional neuroleptic therapy, but the doses that are required are generally greater than those needed for the amelioration of involuntary movements. Problems arise with patients who are psychotic and rigid, as the phenothiazines tend to increase the muscle tone. In each situation the response to drugs must be individually monitored.

9.2. Surgery

The development of the techniques of stereotactic surgery in 1947 heralded a new approach to the treatment of the diseases of the basal ganglia. It is mainly of historical interest to mention that this neurosurgical technique was first used on patients with Huntington's chorea by Spiegel and Wycis in 1949. They performed a small selective lesion in the pallidum of six affected persons (Spiegel and Wycis 1950). The rationale for this method was that choreiform movements may be the result of pallidal overactivity consequent to striatal atrophy, and that therefore pallidal section could possibly abolish this symptom. Three of their six patients had a decrease in movements, two showed no improvement and one person developed a hemiplegia. There are no indications for stereotactic surgery in Huntington's chorea at the present time, as pharmacological agents can now produce the same effects with less serious complications.

In the context of surgery a brief review of anaesthesia in Huntington's chorea is warranted. There have been few reports of anaesthetic management of patients with this disease. Davies (1966) has suggested that such persons may be abnormally sensitive to barbiturates, whilst Gualandi and Bonfanti (1968) and Lamont (1979) have reported a possible sensitivity to neuromuscular depolarising agents. Successful anaesthetic administration to two patients by pancuronium with nitrous-oxide-narcotic and nitrous-oxide-halothane techniques have been described (Farina and Rauscher 1977). It is not clearly established whether the reported abnormalities were idiosyncratic findings in a particular person, who incidentally was affected with Huntington's chorea, or whether these are complications of the disease itself. Further reports on anaesthetic experience in Huntington's chorea are warranted.

9.3. Other Therapeutic Modalities

The role of managing patients with Huntington's chorea and their families requires a combination of specific clinical skills, together with a broad innovative approach. No one therapeutic service applies to all persons with this disease and the clinician can draw flexibly for his specific problem on a whole range of therapeutic modalities which will be reviewed in this section. The general principle underlying utilisation of these services is to use the patient's residual abilities whilst compensating, where possible, for those that have been lost. Most of these services will be especially applicable during the earlier phases of the illness.

In areas where these services are not freely available, the family practitioner will be primarily responsible for the comprehensive care and rehabilitation of the affected person and his family.

9.3.1. Physical Therapy

Physical therapy can prove valuable in prolonging and extending the functional lives of persons with Huntington's chorea (Binswanger 1974). Appropriate exercises may result in an improvement in the patient's energy level and better coordination and balance. Exercise is an important factor in maintaining function at an optimal level and promotes a feeling of well-being, whilst a sedentary existence encourages apathy and promotes muscle weakness.

9.3.2. Speech Therapy

Disturbance of speech with difficulties in communication is one of the earliest and most frustrating aspects of this disease. Speech therapy will help maintain the patient's speech as well as develop alternative ways of communicating when speech becomes unintelligible (Kelly 1977). In the early phases, a slowing down of the speed of speech may help. However, as this form of communication deteriorates, other systems need to be devised so that the patient may still be understood. The establishment of definite signals denoting 'yes' and 'no' is sometimes useful. A chart with space for pictures illustrating such things as 'toilet', 'cigarretes', 'radio too loud' or 'I want a shave' are examples of useful adjuncts.

9.3.3. Occupational Therapy

Patients can be encouraged to use their residual abilities in suitable creative pursuits, such as arts and crafts and other hobbies, which will improve their self-image and restore lost dignity. Borgelt and Linde (1974) have used occupational therapy to help the family and patient maintain a positive attitude and reduce the fearsome aspects of this disease. They found that their patients particularly enjoyed weaving and fitting together ceramic tiles. Heavy nylon was used to string the loom so that excessive tugging or jerking would not break the wool, whilst tiles of at least 1-1½ inches in dimension were used in their ceramic work. Where possible, persons with this disorder should also be included in family outings, picnics or visits to different sports events.

The scope of occupational therapy can be broadened to include providing help with the activities of daily living. Various aids such as non-slip mats, plates with raised edges and rubber build-ups on utensils will make it easier for the person to feed without help. In the bathroom, a bath seat together with hand rails may enable the affected person to wash. Substitutes for buttons and zips will avoid problems with dressing. Numerous other simple, practical devices can foster independence and improve the quality of the patient's life.

9.3.4. Nutrition

Still (1977) has proposed a plan of nutritional therapy in Huntington's chorea, based on the premise that there are similarities between this disease and pellagra. This proposal is not generally accepted and thus far there have been no reports of amelioration of patients' symptoms as a result of any specific diet therapy. Nevertheless, it is obvious that patients with Huntington's chorea should have a well-balanced diet which caters for their own particular needs and fulfils daily requirements of protein, calories and vitamins. Affected persons generally have excellent appetites and supplemental food intake may be necessary to offset the energy loss due to excessive movement.

Dysphagia is a major problem in the late phases of the illness and feeding may be interrupted by choking and coughing episodes. To facilitate swallowing, solids can be cut into small pieces and, if necessary, may be liquidised or softened. Drinking directly from a cup may be extremely difficult and will be helped by the use of a straw.

9.3.5. Nursing Care

The main burden of providing daily care for persons with Huntington's chorea falls on the spouse, who faces both subtle and explicit demands as he or she endeavours to cater for the needs of their partner. These duties range from the preparation of food, bathing

Table 9.3. Nursing care plan

1. Admission assessment

   Vital signs
   BP, temp., pulse, respiration

   Document daily for 7 days after admission, then weekly if normal

   Weight

   Weigh on admission then monthly, or weekly if weight loss occurs

   Skin condition

   Document any bruises, abrasions, dryness

   Motor activity
   Gait, voluntary and involuntary movements

   Note the degree of balance, ambulation, and ability to convey needs

2. Dealing with complications/features

   Difficulty in ambulation
   Loss of balance, ataxic gait

   Assist with ambulation. Have patient use wheelchair and other aids when unable to walk without assistance

   Difficulty in swallowing

   Provide diet as tolerated, from regular to tube feedings. Give liquids through straw. Feed slowly with patient sitting upright

   Choking

   Observe closely to evaluate tolerance of diet. Feed slowly. Do Heimlich's manoeuvre if necessary

   Aspiration

   Clear airways, have suction equipment readily available

   Increase in involuntary movements
   Self-injury, inability to maintain normal body position

   Pad side rails and chairs. Use leg restraints for protection. Do not restrict movements. Observe and reposition frequently

   Inability to perform activities of daily living

   Encourage independence for as long as possible

   Friction burns and abrasions

   Bathe with emollient cleanser; apply lotion. Keep skin clean and dry

   Smoking hazard

   Supervise smoking closely and use smoking aids

   Urinary incontinence

   Apply external catheter (indwelling catheters cause bladder and urethral lacerations due to trauma from excessive or sudden involuntary movements)

   Faecal impaction
   Abnormal bowel pattern, elevated temperature

   Maintain adequate fluid intake: 3000 ml/24 hours unless contraindicated

   Persistent weight loss

   Weigh weekly. Increase caloric intake through diet and supplemental feedings

   Depression

   Encourage patient to describe feelings

   Suicidal intent

   Document and report to physician

   Cough/temperature

   Feed slowly. Report to physician. Check temperature, pulse and respiration until normal for 48 hours

Table 9.3. (continued)

3. Preparation for discharge

General health Note skin condition, including any bruises or abrasions	Instruct patient and family about skin care and positioning: bathe with emollient cleanser, keep skin dry, apply lotion, pay special attention to bony prominences
	Protect patient from hard surfaces in the immediate environment
Activity	If patient unable to walk, instruct patient and family in assisted transfer from bed to chair or bed to commode, and use of walker, cane or orthopaedic device
Bladder control	If incontinent, teach male patient and family use of external catheters
Bowel control	Instruct patient and family to establish a regular time to attempt daily evacuation
Smoking hazard	Caution family to supervise smoking closely and use aids
Medicine	Teach medication schedules, actions and possible side effects
Follow-up care	Explain schedule of visits

Adapted from Stipe et al. (1979).

the affected person and changing their clothes and linen, to helping them come to terms with their illness. Enlisting the aid of district nurses or home helps, when these are available, provides a much-needed respite.

In the early phases of the illness long-term hospitalisation is seldom necessary. However, as the disease progresses, and without adequate social and community resources, long-term hospitalisation becomes an inevitable and necessary eventuality. When to use home care and when to institutionalise patients with Huntington's chorea are difficult decisions which must be made together with the families concerned.

The person with advanced Huntington's chorea has special nursing requirements which need to be considered (Stipe et al. 1979). A nursing care plan is shown in Table 9.3.

9.3.6. Social Work

Social workers are an important part of the health care team for this disease. Advice concerning available community resources, possibilities for rehousing and the means of obtaining financial assistance may greatly alleviate the financial burden. Furthermore, social workers play a crucial role in dealing with emotional problems and resolving family crises (Miller 1976).

9.3.7. Psychotherapy

Psychotherapy can make a major contribution to the improved management of families with Huntington's chorea. In many instances this will not necessarily require the expert attention of a psychiatrist or psychologist, but a practitioner with empathy and insight who has knowledge of this disease and the appropriate referral sources. Requests for advice concerning such issues as driving and employment are frequent. In general it can be stated that patients with Huntington's chorea should not drive, nor should they handle complex machinery.

Other forms of psychotherapy besides those already discussed in Chaps. 7 and 8 may be particularly useful. In many instances affected families will greatly benefit from sharing with persons in similar situations. Group therapy sessions for the unaffected spouse can be a major source of comfort. In these groups persons may share advice on practical issues, which the counsellor, who is not living with the disease, would not be able to provide. These self-help groups foster self-reliance and are an important adjunct to medical management.

9.3.8. Lay Groups

The first lay group organisation was started in the USA in 1967 by Marjorie Guthrie, the widow of Woody Guthrie, the celebrated folk singer who died of Huntington's chorea in that same year. Similar lay groups were established in Britain in 1971 and more recently in Australia, Canada, Belgium, West Germany, the Netherlands, France, Italy and New Zealand. The primary aim of these organisations has been to give help, advice and moral support to affected families.

In the United States there are now several lay groups, including the Committee to Combat Huntington's Disease and the National Huntington's Disease Association, who are primarily directed towards providing services for affected families, and the Huntington's Chorea Foundation and the Hereditary Disease Foundation, whose main thrust is to generate financial support for research into Huntington's chorea. It is noteworthy that a survey in America found that for the majority of persons in affected families the lay organisations were the best source of information regarding the disease (Stern and Eldridge 1975) and that these organisations were as effective as any other source in conveying the risks of transmission of Huntington's chorea in that country. In contrast, however, a recent report has shown that in Britain consultants most commonly inform patients about the disorder (Barette and Marsden 1979).

Lay groups have been successful in drawing the attention of those in government to the needs of persons with Huntington's chorea. One of the outstanding achievements in this area was the establishment of the Commission for the Control of Huntington's Disease and its Consequences in the USA in 1977.

A major area of concern to these organisations and the medical and paramedical professions is the unsatisfactory facilities for residential care for affected patients. These persons are too young for geriatric wards and would also be misplaced in mental institutions. A recent innovation has been the opening of a holiday home in England for affected persons, which also affords necessary respite for the unaffected spouse. The first home for continued care of patients with Huntington's chorea has recently been established in Melbourne, Australia.

Lay organisations play a most important role in the management of Huntington's chorea. Whilst patients and relatives are provided with a framework into which their energies can be directed, doctors in turn can also learn from contact with these groups. The names and addresses of the lay organisations in different parts of the world are listed in Appendix 4. At the Eighth International Workshop on Huntington's Chorea in 1979, a milestone was reached with the formation of an international Huntington's group. The address of the current Secretary-General from the Netherlands is also shown in Appendix 4.

9.4. Future Prospects

Despite the fact that lay organisations have done much to alleviate suffering and improve the care of affected persons and their families, these groups cannot provide an alternative to health care funded at a national level. A more integrated and effective service is needed to help these families. At the same time, a programme of education for medical students, doctors and other professionals, the public and affected families will raise awareness concerning this disorder and prevent those problems which arise from misinformation and ignorance. Encouragement of basic research which may lead to improved care and more effective pharmacotherapy is another urgent need.

The management of patients with Huntington's chorea and their families is a most challenging task. Despite the fact that no cure is available or imminent, appropriate pharmacotherapy and utilisation of other services will improve the care and the quality of life for affected persons and their families. This chapter opened with Huntington's original statement that "no treatment seems to be of any avail" in this disorder. The small but significant advances in the management of this condition justify modification of his original observation.

References

Agid Y (1979) Communication to Eighth International Workshop on Huntington's Chorea, Oxford, England

Albano C, Cocito L (1979) Huntington's chorea and bromocryptine (letter). Arch Neurol 36: 322

Anden NE, Dalen P, Johansson B (1973) Baclofen and lithium in Huntington's chorea. Lancet II: 93

Aquilonius SM, Eckernas SA (1977) Choline therapy in Huntington's chorea. Neurology 27: 887-889

Bachman DS, Butler IJ, McKhann GM (1977) Long-term treatment of juvenile Huntington's chorea with dipropylacetic acid. Neurology (Minneap) 27: 193-197

Barbeau A (1973) Biochemistry of Huntington's chorea. In: Barbeau A, Chase TN, Paulson GW (eds) Huntington's chorea, 1872-1972. Raven Press, New York, pp 525-531

Barette J, Marsden CD (1979) Attitudes of families to some aspects of Huntington's chorea. Psychol Med 9: 327-336

Barr AN, Heinze W, Mendoza JE, Perlik S (1978) Long-term treatment of Huntington's disease with L-glutamate and pyridoxine. Neurology (Minneap) 28: 1280-1282

Binswanger C (1974) Physical therapy in Huntington's chorea. In: Huntington's disease handbook for health professionals. CCHD, New York, pp 17-18

Borgelt PH, Linde TF (1974) Huntington's disease. In: Huntington's disease handbook for health professionals. CCHD, New York, pp 15-16

Brown WT, Sanberg PR, McGeer PL (1979) Corticosteroids and chorea. Arch Neurol 7: 452-453

Calne DB, Teychenne PF, Cleveria LE, Eastman R, Greenacre JK, Petrie A (1974) Bromocryptine in Parkinsonism. Br Med J IV: 442-444

Carlsson, A (1976) Some aspects of dopamine in the basal ganglia. In: Yahr MD (ed) The basal ganglia. Raven Press, New York, pp 181-189

Chase TN (1976) Rational approaches to the pharmacotherapy of chorea. In: Yahr MD (ed) The basal ganglia. Raven Press, New York, pp 337-347

Chiu, E (1979) Notes on the management of Huntington's disease. Aust Fam Physician 8: 197-200

Corrodi H, Fuxe K, Hokfelt T, Lidbrink P, Ungerstedt U (1973) Effect of ergot drugs on central catecholamine neurons: evidence for a stimulation of central dopamine neurons. J Pharm Pharmacol 25: 409-413

Corsini GU, Onali P, Masala C, Cianchetti C, Mangoni A, Gessa G (1978) Apomorphine hydrochloride-induced improvement in Huntington's chorea: stimulation of dopamine receptor. Arch Neurol 35: 27-30

Dalby MA (1969) Effect of tetrabenzine on extrapyramidal movement disorders. Br Med J ii: 422-423

Davies DD (1966) Abnormal response to anaesthesia in a case of Huntington's chorea. Br J Anaesth 38: 490-491

Enna SJ, Maggi A (1979) Biochemical pharmacology of gabaergic agonists. Life Sci 24: 1727-1738

Fahn S (1972) Treatment of choreic movements with perphenazine. Dis Nerv Syst 33: 653-658

Fahn S (1973) Treatment of choreic movements with perphenazine. In: Barbeau A, Chase TN, Paulson GW (eds) Huntington's chorea, 1872-1972. Raven Press, New York, pp 755-764

Farina J, Rauscher LA (1977) Anaesthesia and Huntington's chorea: a report of two cases. Br J Anaesth 11: 1167-1168

Frattola L, Albiazzati MG, Spano PF, Trabucchi M (1977) Treatment of Huntington's chorea with bromocriptine. Acta Neurol Scand 56: 37-45

Growden JH, Cohen EL, Wurtman RJ (1977) Huntington's disease: clinical and chemical effects of choline administration. Am Neurol 1: 418-422

Gualandi W, Bonfanti G (1968) Un caso di apnea prolungata in corea di Huntington. Acta Anaesth (Padua) 19: 235-238

Horn, AS, Snyder SH (1972) Chlorpromazine and dopamine conformational similarities that correlate with the antischizophrenic activity of phenothiazine drugs. Proc Natl Acad Sci USA 68: 2325-2328

Jenner P, Marsden CD (1979) The substituted benzamides: a novel class of dopamine antagonists. Life Sci 25: 479-486

Kartzinel R, Perlow MD, Carter AC, Chase TN, Calne DB, Shoulson I (1976) Metabolic studies with bromocriptine in patients with idiopathic Parkinsonism and Huntington's chorea. Trans Am Neurol Assoc 101: 53-56

Kelly R (1977) Testimony to Commission for Control of Huntington's Disease and its Consequences, vol IV/2, US Government Printing Office, Washington, DC, pp 77-78

Kingston D (1979) Tetrabenzine for involuntary movement disorders. Med J Aust I: 628-630

Klawans HL, Paulson GW, Ringel SP, Barbeau A (1972) The use of L-dopa in the detection of presymptomatic Huntington's chorea. N Engl J Med 286: 1332-1334

Lamont AM (1979) Brief report: anaesthesia and Huntington's chorea. Anaesth Intens Care 7: 189-190

Lazarte JA, Petersen MC, Baars CW, Pearson JS (1955) Huntington's chorea: results of treatment with reserpine. Proc Mayo Clin 30: 358-365

Loeb C, Roccatagliata G, Albano C, Besio G (1979) Bromocryptine and dopaminergic function in Huntington's disease. Neurology (Minneap) 29: 730-734

Miller E (1976) The social work component in community-based action on behalf of victims of Huntington's disease. Soc Work Health Care 76, 2(1): 25-32

Perry TL, Hansen S, Klosker M (1973) Huntington's chorea: Deficiency of γ-aminobutyric acid in brain. N Engl J Med 288: 337-342

Perry TL, Wright JM, Hansen S, MacLeod P (1979) Isoniazid therapy of Huntington's chorea. Neurology 29: 370-377

Perry TL, Wright JM, Hansen S, Allen BM, Baird PA, MacLeod PM (1980) Failure of aminooxyacetic acid therapy in Huntington's disease. Neurology (Minneap) 30: 772-775

Ringel SP, Guthrie M, Klawans HL (1973) Current treatment of Huntington's chorea. In: Barbeau A, Chase TN, Paulson GW (eds) Huntington's chorea, 1872-1972. Raven Press, New York, pp 797-801

Schou M (1979) Lithium in the treatment of other psychiatric and non-psychiatric disorders. Arch Gen Psychiatry 36: 856-858

Shoulson I (1977) Clinical care of the patient and family with Huntington's disease. In: Commission for the Control of Huntington's Disease and its Consequences, vol II. US Government Printing Office, Washington, DC, pp 421-451

Shoulson I (1979) Huntington's disease: overview of experimental therapeutics. In: Chase TN, Wexler NS, Barbeau A (eds) Advances in neurology, vol 23. Raven Press, New York, pp 751-759

Shoulson I, Chase TN, Roberts E, Van Balgooy JN (1975) Huntington's disease: treatment with imidazole-4-acetic acid. N Engl J Med 293: 504-505

Shoulson I, Kartzinel R, Chase TN (1976) Huntington's disease: treatment with dipropylacetic acid and gamma-aminobutyric acid. Neurology (Minneap) 26: 61-63

Shoulson I, Goldblatt D, Charlton M, Joynt RJ (1978) Huntington's disease: treatment with muscimol, a GABA-mimetic drug. Ann Neurol 4: 279-284

Spiegel EA, Wycis HT (1950) Pallido-thalamotomy in chorea. Arch Neurol Psychiatr 64: 295-296

Stern R, Eldridge R (1975) Attitudes of patients and their relatives to Huntington's disease. J Med Genet 12: 217-223

Still CN (1979) Nutritional therapy in Huntington's chorea: concepts based on the model of pellagra. Psychiatr Forum, 74-78

Stipe J, White D, Van Arsdale E (1979) Huntington's disease. Am J Nurs 79(8): 1428-1432

Stokes HB (1975) Trifluoperazine for the symptomatic treatment of chorea. Dis Nerv Syst 36: 102-105

Taglia JU, McGlamery M, Sambandham RR (1978) Tetrabenzine in the treatment of Huntington's chorea and other hyperkinetic movement disorders. J Clin Psychiatr 39: 81-87

Tarsy D, Leopold N, Sax DS (1974) Physostigmine in choreiform movement disorders. Neurology (Minneap) 24: 28-33

Tolosa ES, Sparber SB (1974) Apomorphine in Huntington's chorea: clinical observations and theoretical considerations. Life Sci 15: 1371-1380
Whittier JR, Korenyi C (1968) Effect of oral fluphenazine on Huntington's chorea. Int J Neuropsychiatr 4: 1-3
Zanglein JP, Buldauf E, Hussaini M (1978) A new treatment of Huntington's chorea: report of three cases. Sem Hop Paris 54: 871-874

10 Current Trends in Research

The number of articles on Huntington's chorea in the scientific literature has increased dramatically over the past 10 years. In 1970 there were 27 citations on Huntington's chorea in *Index Medicus* and this had trebled to 83 in 1979. Another expression of the recent upsurge in interest was the establishment of the Commission for the Control of Huntington's Disease and its Consequences in the USA in 1977, whose purpose was to report on the state of research and management in this disorder. In October 1977 their report was tabled before the Congress and President of America and as a result of their recommendations there has been added financial support for research on Huntington's chorea in America and other parts of the world.

Considerable attention was devoted in the early 1970s to the post-mortem investigation of affected persons' brains, in an effort to define the characteristic neurochemical abnormality of Huntington's chorea. This approach was pursued following the elucidation of the role of dopamine deficiency and the advent of subsequent successful dopamine agonist therapy in Parkinson's disease. It was hoped that the exploration of neurotransmitter dysfunction in Huntington's chorea might lead to improved therapy and possibly an understanding of the basic defect in this disease. However, the early promise has not been fulfilled and the focus of research activity has now shifted from neurochemistry to other areas, particularly those of cell biology and membrane physiology.

In this chapter a brief résumé of the state of research into Huntington's chorea will be presented. It is not intended to be a comprehensive review, but emphasis will be placed on the most promising recent developments. Readers desiring more extensive information are referred to the excellent articles on this subject to be found in *Advances in Neurology*, vol 23 (1979).

10.1. A Unifying Conceptual Approach

A paradigm for the development of the abnormal findings in Huntington's chorea is shown in Fig. 10.1. This model provides an overview of the different lines of research that are being pursued in search of the basic defect in this disorder and forms the framework for the presentation of this chapter.

The past few years have witnessed progress in the understanding of the secondary effects of the abnormal gene without comparable advances in the elucidation of the basic underlying mechanisms of this disease. The search for the altered gene product is, however, now being intensified as new research techniques are being developed. A major conceptual advance has been the recognition that even though Huntington's chorea is

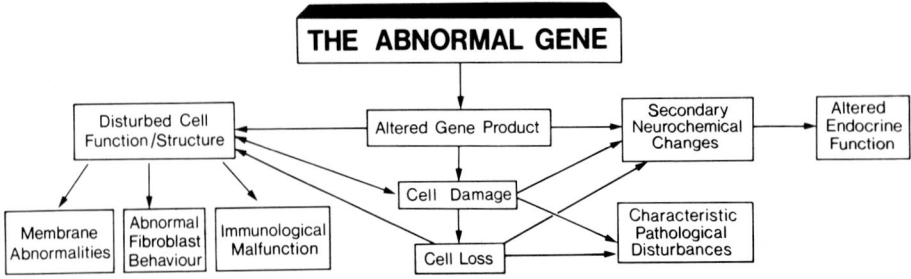

Fig.10.1. A paradigm for the development of the abnormal findings in Huntington's chorea (Adapted from Barbeau 1977, and Stevens 1980)

essentially a disorder of the central nervous system, the primary defect may be expressed in cells outside the brain, as manifest by disturbances in extraneural cell function and structure. There is, in fact, accumulating evidence from studies of erythrocytes, lymphocytes and fibroblasts of affected persons which indicates that the basic defect may be measurable in these easily obtainable tissues and that Huntington's chorea is more generalised than was initially suspected. The finding of a consistent abnormality in an easily sampled tissue outside the nervous system would, by virtue of its accessibility to research endeavour, be of considerable value in hastening the search for the basic defect in the disorder.

At the present time it is uncertain which area of research will be the most fruitful and each of the varied disciplines described below has in some way enhanced the understanding of the disorder.

10.2. Investigations of the Abnormal Gene

The gene for Huntington's chorea lies on one of the autosomal chromosomes, but its exact location remains unknown. There are two principal methods for gene localisation, only one of which has been attempted in Huntington's chorea. The classic approach to gene mapping is based on the principle that genes lying in close proximity to each other tend to be inherited together and are regarded as being 'linked', whilst genes far apart segregate independently. To date there have been several attempts to determine the linkage relationships of the gene for Huntington's chorea, but none has been conclusive.

Pericak-Vance et al. (1979) recently examined the linkage of the abnormal gene to 27 polymorphic marker systems and found no significant linkage. Went and Volkers (1979) also reported negative findings in similar experiments.

The second method for gene localisation is based on the principle of somatic cell aggregation in cell hybrids, usually mouse/human. As these cells divide chromosomes are randomly lost, and when loss of a chromosome is correlated with loss of a particular enzyme function it can be inferred that the gene for that enzyme is located on that particular chromosome. Unfortunately this technique has not been used in Huntington's chorea, as the prerequisite of the method is knowledge of the biochemical effect of the abnormal gene.

In spite of numerous investigations conventional linkage studies have been largely unrewarding. However, the recent development of recombinant DNA methodology offers a novel approach for linkage experiments in this disease. These methods greatly increase the number of genetic markers available for linkage studies and will facilitate chromosomal assignment of the abnormal gene. The most obvious application of such knowledge would be prenatal diagnosis of this disease. However this would also shed light on the unanswered problem of heterogeneity and could theoretically aid in the elucidation of the molecular defect in Huntington's chorea.

10.3. The Search for the Altered Gene Product

The fundamental unanswered question in Huntington's chorea is why certain cells selectively die at a particular time, and the clue to this problem may be found in the fact that Huntington's chorea is a dominantly inherited genetic disease. Whilst most autosomal recessive diseases involve enzyme defects, a reasonable generalisation is that most autosomal dominant disorders involve structural proteins. Until 1974 there were no reports of protein abnormalities in Huntington's chorea. However, in that year Igbal et al. reported the presence of abnormal concentrations of three basic proteins in the neuronal microsomal fraction, while Kjellin and Stibler (1974) performed electrophoresis of cerebrospinal fluid (CSF) from affected patients and found abnormal alkaline-end fractions in the CSF gamma-globulin region. These were, however, not specific to Huntington's chorea.

The two-dimensional gel electrophoresis technique is a particularly useful method for studying normal proteins that separates them on the basis of charge and molecular weight. Heterozygotes for Huntington's chorea have one faulty gene for the disease and one normal allele. Theoretically it could be expected that the sample from the affected person would yield two spots, representing the normal and abnormal proteins, as opposed to the single spot seen in controls. Comings' (1979) search for the mutant protein based on this rationale has thus far been negative. Perry (1979) has used similar techniques and found a substance similar in structure to the neurotoxin kainic acid in serum from patients with Huntington's chorea, but not in control serum. The exact composition of this factor is

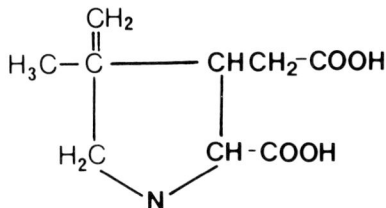

Fig.10.2. The structural similarity between kainic acid and glutamic acid

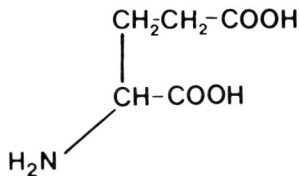

Table 10.1.   The kainic acid (KA) model for Huntington's chorea

Similarities to Huntington's chorea	Differences from Huntington's chorea
Neurochemistry	
Decreased striatal GABA, glutamic acid decarb-oxylase (GAD), choline acetyltransferase (CAT), angiotensin converting enzyme (ACE) and encephalin concentration	
Normal striatal dopamine, serotonin and tyrosine hydroxlase concentrations. Decreased GABA, GAD and substance P in the substantia nigra	
Receptors	
Decreased muscarinic and serotonin receptors	Marked decrease in GABA receptor binding after KA injection
Pathology	
Loss of striatal neurones with ventricular dila-tation. Intense gliosis	KA-induced neuronal loss generally more complete and involves areas usually spared in Huntington's chorea
Clinical	
Decrease in body weight. Impaired learning and defective memory	More acute time course, cells dying rapidly. Dissimilar motor symptoms
Genetics	
	KA effects non-genetic

still to be determined. In view of its structural similarity to kainic acid it is feasible that this substance acts as a toxin, causing selective cell death in the brain of patients with Huntington's chorea. In this context a brief review of the effects of kainic acid in rat brain and a comparison of these effects with Huntington's chorea is warranted.

Kainic acid (KA), a glutamate analogue (Fig. 10.2), when injected into rat striatum promptly results in various pathological neurochemical and clinical changes which are very similar to those associated with Huntington's chorea. KA-induced striatal lesions have thus been proposed as a useful animal model for Huntington's chorea. Certain differences exist however (Table 10.1), which preclude complete extrapolation from this model to affected patients (McGeer et al. 1979; Marsden 1979; Olney 1979; Sanberg and Fibiger 1979).

Animal models of Huntington's chorea may be extremely important tools even if they do not completely parallel the human disorder. The KA-induced experimental model provides a means for the testing of new drugs of possible therapeutic benefit in Huntington's chorea. A better understanding of the mechanism of neurotoxicity of kainic acid may also shed light on the cause of neuronal destruction in this disease. Kainic acid exerts its effects only in the presence of an intact corticostriatal glutamate pathway (McGeer et al. 1977) and the neurotoxic effect is thought to be due to its excitatory action at the glutamate receptor, which is damaged as a result of excessive stimulation (Olney and De Gubareff 1978). This hypothesis, however, is unproven and the mechanisms of cell death in both the KA-induced lesions and Huntington's chorea remain undetermined.

Olney (1979) has explored the possible toxic role of glutamic acid in the causation of Huntington's chorea. In this context it is relevant that glutamine is toxic to fibroblasts of affected individuals but has no detrimental effect on those of healthy persons. A further application of the kainic acid animal model could be the successful development of specific glutamate antagonists, which may be of benefit in the treatment of patients with Huntington's chorea (Marsden 1979).

10.4. The Investigation of Disturbed Cell Function/Structure

The abnormal gene causing Huntington's chorea is present in every cell of the body even though the brain is the principal site of pathology. Obvious difficulties associated with the study of brain tissue during life have prompted workers to look outside the nervous system for a more accessible tissue which might still show some subtle yet specific alteration in cell behaviour or structure.

Red blood cells, platelets, fibroblasts and lymphocytes of patients with Huntington's chorea have been subjected to complex investigative techniques and numerous abnormalities have been found.

10.4.1. Membrane Abnormalities

Current investigations of erythrocytes, fibroblasts and neurones point to a generalised disturbance in membrane function and structure in Huntington's chorea.

Butterfield and his colleagues (1977) have employed biophysical techniques involving electron spin resonance measurements and found impaired red cell membrane fluidity, which could be partially reversed by preincubation with GABA. Scanning electron microscopy was used by the same workers (Markesberry and Butterfield 1977), who found an increased number of stomatocytes in Huntington's chorea and suggested that this resulted from a disturbed response to fixation by an altered erythrocyte membrane. An increase in the activity of sodium/potassium-stimulated ATPase in red cell membrane in this disorder is further evidence for a membrane abnormality.

Fluorescence spectroscopy is a sensitive technique used to monitor dynamic changes in membranes. Using a fluorescent probe specific for membrane polar–apolar interfaces, Pettegrew et al. (1979) have reported marked differences at low passages in excitation, emission and polarisation spectra between fibroblasts of affected persons and controls. This adds support to the concept of a membrane abnormality in Huntington's chorea.

Lymphocyte surface-receptors become cross-linked on exposure to an appropriate ligand and subsequently redistribute to one pole of the cell to form a cap. Noronha et al. (1979) have shown that lymphocytes of persons with Huntington's chorea show decreased capping, which may point to a lymphocyte membrane abnormality in the disease.

Gangliosides are constituents of cell membranes and are found in highest concentrations in the nervous system, predominantly in neurones. The ganglioside concentration is significantly decreased in the caudate nucleus and putamen of patients with Huntington's chorea, which in addition has an altered ganglioside pattern (Bernheimer et al. 1979).

The findings of all these recent investigations provide evidence of a generalised abnormality in cell membrane structure and function in Huntington's chorea (Table 10.2). Whether these changes are in part the effects of neuroleptic or other drug administration in Huntington's chorea has not been assessed. Furthermore, the postulated membrane abnormalities still fail to provide adequate explanations for the well-documented pathological and neurochemical findings in this disease.

Table 10.2. Evidence for membrane abnormalities in Huntington's chorea

System	Evidence	Author(s)
Red blood cells	Altered electron spin resonance	Butterfield et al. (1977)
	Altered morphology, increased stomatocytes	Markesberry and Butterfield (1977)
	Increased Na-K-ATPase activity	Butterfield et al. (1978a)
	Decreased deformability	Butterfield et al. (1978b)
Fibroblasts	Disturbed fluorescence spectroscopy	Pettegrew et al. (1979)
Lymphocytes	Decreased capping	Noronha et al. (1979)
Neurones	Decreased and altered ganglioside concentration	Bernheimer et al. (1979)

10.4.2. Fibroblast Activity

Fibroblasts in culture are particularly useful tools for the study of metabolic and morphological changes in human cells. For research on Huntington's chorea they provide a proliferating, dynamic system which can be challenged by selective depletion or supplementation of the nutritional environment in an attempt to define and then correct the inherited defect.

Since Menkes' and Stein's report in 1973 of a decreased lifespan and possible accelerated ageing of cultured fibroblasts from patients with Huntington's chorea, there have been numerous other reports of abnormal fibroblast behaviour in this condition (Table 10.3). Some workers have reported a normal initial growth rate for these

Table 10.3. Abnormal fibroblast behaviour in Huntington's chorea

Growth characteristics of Huntington's chorea fibroblasts compared with controls	Evidence for	Evidence against
Decreased replicative lifespan	Menkes and Stein (1973)	Goetz et al. (1975) Gray and Dana (1979)
Grow to higher confluent density	Goetz et al. (1975) Kirk et al. (1977) Barkley et al. (1977) Leonardi et al. (1978) Goetz et al. (1979)	Brown et al. (1979) Cassiman et al. (1979) Beverstock (1979) Gray and Dana (1979)
Grow to higher density in early but not later passages. Have glucosamine dependence for normal growth	Tourian and Hung (1977a)	
Unusual sensitivity to effects of glutamine	Tourian and Hung (1979a)	
Impaired adhesion	Tourian and Hung (1977b)	
Significantly enhanced radiosensitivity	Arlett and Muriel (1979)	Arlett (1980)
Synthesise ×3 the amount of GABA	Grey and Dana (1979)	

fibroblasts but later growth to a higher confluent density than controls. More recent reports have not been able to confirm these findings (Table 10.3). Other abnormal features which have been documented include glucosamine dependence for normal growth, impaired adhesion to a plastic substratum and undue sensitivity to the toxic effects of glutamine. This toxic effect is characterised by cell elongation and cell death within 2–3 weeks.

Recently fibroblasts from four affected persons have been shown to be abnormally sensitive to ionising radiation (Arlett and Muriel 1979); this radiosensitivity has also been shown in the study by Moshell et al. (1980) of cultured lymphocyte lines in the disorder. However, following the refutation by Arlett (1980) of his own initial findings, doubts have been cast on the significance of this investigative method.

A major problem in the interpretation of all these findings is that different workers have often produced conflicting results, which has led some investigators to question the value of this approach in Huntington's chorea. Factors influencing fibroblast growth that must be carefully considered include the site of biopsy, the method of dissection and the age of the donor. For example, papillary fibroblasts exhibit greater proliferative capacities in vitro than reticular fibroblasts from the same skin specimen (Harper and Grove 1979), as do cells from young donors compared with those from older persons (Schneider 1979). Furthermore, cells biopsied from the forearm grow faster than those from the upper arm (Brown et al. 1979), and fibroblast growth patterns may also be altered by differences in the composition of the nutrient medium and in freezing and storage (Goetz et al. 1979). These may be some of the reasons for the disagreement in results of different studies. Nevertheless, it would seem that some discrepancies in growth characteristics of skin fibroblasts from patients and from controls do exist; their possible significance awaits further confirmation.

10.4.3. Immunological Abnormalities

The involvement of the immune system in the pathogenesis of Huntington's chorea was first suggested by McHenemey in 1961 and more recently has been considered by several investigators who have found both cellular and humoral immune abnormalities (Table 10.4).

Barkley et al. (1977, 1978) have provided evidence for a cellular immune response in Huntington's chorea. They demonstrated an inhibition of migration of patients'

Table 10.4. Evidence for immunological abnormalities in Huntington's chorea

Evidence	Author(s)
Cellular immune responses: affected lymphocytes produce migration inhibitory factor when confronted with Huntington's chorea and multiple sclerosis brain extracts	Barkley et al. (1977, 1978)
Alterations in immediate hypersensitivity responses: presence of antineuronal antibody unlike that seen in controls	Husby et al. (1977, 1979a)
Decreased immunological activity of lymphocytes: impaired ability to effect antibody-dependent killing	Morrell (1979)

lymphocytes on exposure to brain of persons with Huntington's chorea, in contrast to their normal progression when confronted with normal brain extracts. This finding supports the hypothesis that the degenerating brain has altered antigenic properties. If a sequestered brain antigen were released into the circulation it could thus sensitise patients' lymphocytes. Whilst this may be true, it still does not explain what caused the initial cellular damage.

The presence of antineuronal antibodies in affected persons' serum led Husby et al. (1977) to speculate that immunoglobulins may also be involved in the pathogenesis of cellular damage. However, the presence of these antibodies in 30% of unaffected spouses and only 46% of Huntington's chorea subjects suggests that they are not directly involved in the disease process. Also, the presence of antineuronal antibodies in 11 out of 33 persons with Parkinson's disease suggests that the phenomenon of antineuronal antibody reactivity is not unique to Huntington's chorea. Husby et al. (1979) have determined whether the antineuronal antibodies in Huntington's chorea showed any restriction with respect to IgG subclass light-chain types and heavy-chain variable subgroups. Whilst there was a predominance of one light chain of the λ-type there was extensive variability of the heavy-chain subgroups, which further attested to the lack of correlation within the antineuronal antibody population of patients with Huntington's chorea.

Morrell (1979) has demonstrated a significant abnormality in the capacity of lymphocytes from affected persons to effect antibody-dependent killing in an immunological assay. The mechanism and significance of this impairment are at present unknown.

Knowledge of the association between the tissue antigens of the human leucocyte system A (HLA) and different diseases is progressing at a rapid rate. At the present time, however, coupling between specific HLA antigens and Huntington's chorea has not been documented (Foerster and Freudenberg 1980).

10.5. Viruses

The discovery that viruses could cause chronic and fatal neurodegenerative disease has been an impetus for workers to look for evidence which might implicate viruses in the causation of Huntington's chorea. This hypothesis was strengthened by the observation that the slow virus encephalopathy, Creutzfeldt–Jacob disease, could infrequently be inherited as an autosomal dominant trait.

Recently a virus-like agent has been identified in the cerebrospinal fluid of a proportion of persons with schizophrenia (Crow et al. 1979). The possibility that Huntington's chorea could be due to a virus-like agent which promotes neuronal death in a particular genetically predisposed subpopulation has been suggested. The finding of antineuronal antibodies in spouses of affected persons raises the possibility that they may be exposed to a similar infective agent, but do not develop the disease because of their different genetic constitution. The report of focal inflammation with virus-like structures in the brains of patients with Huntington's chorea and not in controls, has given further support to this hypothesis (Auerbach 1980). However, such particles were not found in the classic intranuclear location and the failure to culture a virus has raised some doubt about the significance of this finding. Furthermore, animals inoculated with autopsied brain tissue of persons with Huntington's chorea have not contracted this disease (Gajdusek 1977). Whilst Auerbach's findings are intriguing, there is insufficient evidence firmly to support the theory that Huntington's chorea is due to a virus which is genetically transmitted. Nevertheless this possibility deserves further consideration.

10.6. Neurochemistry

Tremendous progress has been made in the elucidation of the neurochemical defects of Huntington's chorea. Post-mortem analyses of the basal ganglia of affected patients have demonstrated abnormalities in numerous neurotransmitter systems, including dopamine, gamma-aminobutyric acid (GABA), acetylcholine, serotonin, substance P, angiotensin-converting enzyme (ACE), the encephalins, vasoactive intestinal polypeptide (VIP) and cholecystokinin (CCK) (Table 10.5; Fig. 10.3). Whilst these studies have aided in the characterisation of these different neurotransmitters (Fig. 10.4) and the elucidation of their function, they have not, however, significantly enhanced our understanding of the underlying defect of Huntington's chorea. The neurochemical abnormalities of the dopaminergic and non-dopaminergic neurotransmitter system of this disease will now be described.

10.6.1. Dopamine

The neurochemical finding of dopamine deficiency in the nigro-striatal pathways of Parkinsonian patients and the effective utilisation of L-dopa in the amelioration of these persons' symptoms focused attention on dopamine function in Huntington's chorea. Bernheimer et al. (1973) examined dopamine concentration in ten persons with Huntington's chorea and found a slightly reduced concentration in the caudate nucleus, with normal levels in other brain regions. Mattsson (1974) reported no difference in dopamine concentrations from controls in his investigation of five choreic brains. Bird and Iverson (1974) found an insignificant decrease in caudate nucleus dopamine concentration in 17 affected persons' brains, which became statistically significant, however, on examination of a larger patient population (Bird and Iverson 1976). The most recent and comprehensive investigation was performed by Spokes (1979, 1980) on 56 choreic brains, and he found an elevated dopamine concentration in the nigrostriatal and mesolimbic dopamine systems. The relative increase of dopamine in the corpus striatum and pars compacta of the substantia nigra and the nucleus accumbens could, in part, be due to the relative sparing of the nigrostriatal dopaminergic system compared with the more generalised cell loss in other regions.

The inconsistent findings of decreased, normal and increased dopamine concentration in the brains of deceased patients highlight the difficulties, and the caution that must be exercised, in the interpretation of these results. Ante-mortem and post-mortem factors which may have contributed to these discrepancies include age, prior drug administration, cause of death, poor selection of controls, a small patient sample, the delay between death and tissue storage, and postponement of autopsy examination.

Tyrosine hydroxylase, the biosynthetic enzyme for dopamine and dopamine-beta-hydroxylase, its degradative enzyme, have also been measured in choreic brains, with no significant abnormalities being found (Bird and Iverson 1974).

10.6.2. Gamma-aminobutyric Acid (GABA)

GABA was the first amino acid shown to function as a neurotransmitter in both vertebrate and invertebrate nervous systems and there is strong evidence that it exerts a central inhibitory function (Curtis 1976). Perry and his colleagues (1973) noted a marked reduction of this neurotransmitter in the substantia nigra and corpus striatum of patients with Huntington's chorea. Other investigators (Table 10.5) reported a significant decrease in glutamic acid decarboxylase (GAD) activity, the enzyme converting glutamic

Table 10.5. The neurochemistry of Huntington's chorea: concentrations compared with normal controls

Neurotransmitters, metabolites, peptides	Related enzymes	Caudate	Putamen	Gl. pallidus	S. nigra	Nuc. accumbens	Cortex	CSF serum	Author(s)
Dopamine		N[1,5], ↓[2], ↑[3]	N[2], ↑[3]	N[2], ↑[3]	N[2], ↑[3]	↑[3], ↓[4]			1. Bird and Iverson (1974) 2. Bernheimer et al. (1973) 3. Spokes (1979) 4. Barbeau (1979) 5. Mattsson (1979) 6. Curzon et al. (1972)
	Tyrosine hydroxylase	N[1]	N[1]		↓[1]				
HVA		N[2,5], ↓[6]						↓[2,6]	
	Dopamine-β-hydroxylase	N[1]						N[5]	
GABA		↓[1,2]	↓[1]	↓[1]	↓[1]		↓[1], N[1]	↓[5]	1. Perry et al. (1973) 2. Bird et al. (1973) 3. Bird and Iverson (1974) 4. Wu et al. (1979) 5. Glaeser et al. (1975) 6. Roth et al. (1979) 7. Bird and Iverson (1976)
	Glutamic acid decarboxylase	↓[3]	↓[4]	↓[7]	↓[7]	↓[4], N[7]			
	GABA transaminase		N[4]						
	Gamma-hydroxybutyrate	↓[6]	↓[6]		↓[6]				
Homocarnisine					↓[1]				
Acetylcholine	Choline acetyltransferase	↓	↓	N	N		N		Bird and Iverson (1974)

Neurotransmitters, metabolites, peptides	Related enzymes	Caudate	Putamen	Gl. pallidus	S. nigra	Nuc. accumbens	Cortex	CSF serum	Author(s)
Serotonin		N[1]	N[1]		N[1]				1. Bernheimer et al. (1973) 2. Curzon et al. (1972) 3. Caraceni et al. (1977)
5HIAA								N[2,3]	
Substance P		N	N	↓	↓		N		Kanazawa et al. (1979)
Angiotensin II	ACE	↓	↓	↓	↓	↓	N		Arregui et al. (1977)
Encephalins				↓	↓				Arregui et al. (1979)
Somatostatin				N					Arregui et al. (1979)
VIP		N	N	N	N		N		Arregui et al. (1979)
CCK		N	N	↓	↓		N		Emson et al. (1980)

N, normal; ↓, decreased; ▲, increased.

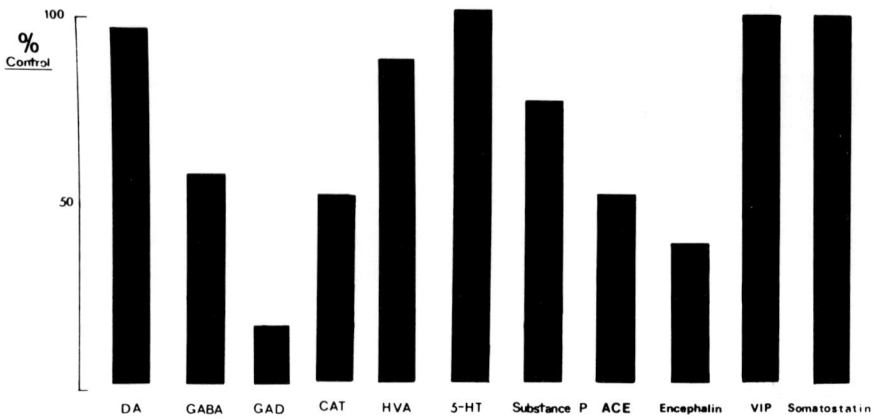

Fig.10.3. A neurochemical profile of the basal ganglia in Huntington's chorea

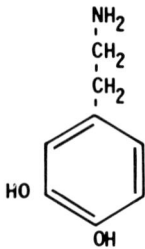

DOPAMINE

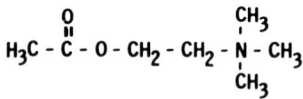

ACETYLCHOLINE

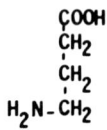

GABA

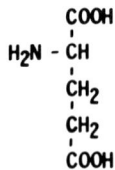

L-GLUTAMIC ACID

Fig.10.4. The chemical structure of neurotransmitters which may have an important role in the pathophysiology of Huntington's chorea

acid to GABA, which is a stable and specific marker for GABA-containing neurones (McGeer et al. 1973). It was hoped that this consistent abnormality might reflect the primary defect specific to Huntington's chorea. Bird (1976) searched for evidence of defective GAD function in non-neural tissue to support this hypothesis, but was not able to demonstrate any difference in GAD function in the kidneys of affected persons compared with controls.

Perry and co-workers (1978) recently showed that GABA is reduced as markedly in the brains of patients with schizophrenia as in the brains of Huntington's chorea patients, thereby refuting the hypothesis that the primary defect in the GAD/GABA system is specifically related to the latter disorder.

The depression in GABA concentration in Huntington's chorea is a secondary effect of an underlying genetic abnormality which is still unknown and is only part of the complex neurochemical disturbance of this disease. The failure to alleviate patients' symptoms by augmenting brain GABA levels highlights the fact that neurotransmitter function is not only directly related to its concentration, but also reflects the net effects of its secretion, turnover, inactivation, interaction with receptors and relative concentrations of antagonistic neurotransmitters.

10.6.3. *Acetylcholine*

The corpus striatum contains high concentrations of acetylcholine and its synthesising enzyme, choline acetyltransferase (CAT). Biochemical evidence of cholinergic neurone involvement in Huntington's chorea derives from the finding of a marked reduction (50%) in the activity of choline acetyltransferase in the striatum and putamen of affected persons (Table 10.5). However, the defect in the cholinergic system is not primary, but rather secondary to abnormalities in other neurochemical systems.

10.6.4. *Serotonin*

Serotonin interacts with dopamine in the control of motor behaviour. However, the finding of normal serotonin concentration in the basal ganglia of patients with Huntington's chorea (Table 10.5), together with the failure of drugs which alter serotonin function to modify patients' symptoms (Ringel et al. 1973), suggest that this system is not primarily affected in this disease.

10.6.5. *Peptides*

The role of peptides in the general functioning of the central nervous system is receiving increased attention at present and their influence on motor control is particularly relevant to Huntington's chorea.

The undecapeptide, substance P, is an excitatory neurotransmitter, found in highest concentration in the substantia nigra of the central nervous system. The selective and specific decrease in substance P in the substantia nigra and globus pallidus of patients with Huntington's chorea (Kanazawa et al. 1977, 1979) has highlighted the possible role of this peptide in extrapyramidal function (Table 10.5).

Substance P stimulates striatal dopamine receptors (Magnusson et al. 1976) whilst GABA acts as an inhibitor of these dopamine neurones (Anden 1974). In other words, substance P and GABA seem to have antagonistic actions in the central nervous system. Diamond et al. (1979) have suggested that even though substance P is decreased in certain areas of the brain of patients with Huntington's chorea, there is an even greater decrease in GABA, resulting in a relative preponderance of substance P. This could be of major therapeutic significance, as drugs which act specifically on substance P systems may improve the treatment of this disease.

Angiotensin-converting enzyme (ACE) is a specific dipeptide which converts the inactive decapeptide angiotension I to the octapeptide angiotensin II. There is increasing evidence that the nigrostriatal neuronal system contains ACE activity, and an 83%–92% reduction in the striatum of patients with Huntington's chorea has been reported (Arregui et al. 1979; Table 10.5). Whilst little is known about the functional role of ACE in the central nervous system, it is noteworthy that patients with Huntington's chorea have defects in postural vasoregulatory mechanisms, possibly implicating abnormalities in the renin–angiotensin system in this disease. In this context it is relevant that plasma renin activity in patients is normal (M. Hayden and P. Sever, 1980, unpublished work).

Two pentapeptide encephalins have been identified in the brain which differ in their amino-terminal amino acids and are named methionine-encephalin and leucine-encephalin respectively. Methionine-encephalin concentration is reduced in the globus pallidus and substantia nigra in Huntington's chorea (Arregui et al. 1979; Table 10.5). Whilst the exact role of the encephalins in the basal ganglia is unknown, it has been proposed that they modulate the turnover of dopamine (Diamond and Borison 1978).

Other neuropeptides which have been investigated include somatostatin (SRIF) and vasoactive intestinal polypeptide (VIP), which are found in similar concentrations in patients with Huntington's chorea and normal controls (Arregui et al. 1979). A reduction in cholecystokinin (CCK)-like immunoreactivity in the globus pallidus and substantia nigra of affected persons has recently been reported (Emson et al. 1980).

10.6.6. Neurotransmitter Receptor Abnormalities

Receptors are characterised by their ability to recognise and respond to minute quantities of neurotransmitters and other substances in some precise way. The clinical manifestations of Huntington's chorea may not only result from altered neurotransmitter concentration, but may also be influenced by disturbances in receptor structure and function. There are precedents for this as receptor mutations are known to be primarily involved in the pathogenesis of other autosomal dominant disorders, including familial hypercholesteraemia (Frederickson et al. 1978) and some forms of haemolytic anaemia (Valentine et al. 1977).

Recent improvements in research techniques have facilitated the investigation of neurotransmitter receptor function in Huntington's chorea. Alterations in cholinergic, serotonin, beta-adrenergic, GABA, dopamine and benzodiazepine receptors have been reported (Table 10.6). Altered receptor binding may result from either a reduction in the number of receptor sites or a decreased affinity for the binding agent. In most reports the reduced binding could be attributed to the existence of fewer receptor sites without a change in the binding properties of remaining receptors. There is some evidence, however, that striatal GABA binding is increased in Huntington's chorea (Lloyd and Davidson 1979).

Weiner et al. (1979) have provided evidence for two separate subpopulations of striatal dopamine receptors which respond differently to pharmacological manipulations. Kebabian and Calne (1979) have further delineated these receptors on the basis of their ability to increase cyclic AMP synthesis via their capacity to stimulate adenyl cyclase (D1 receptors). Those receptors with no effect on cyclic AMP turnover are termed D2 receptors. The concept of multiple categories of dopamine receptors may have important implications for the pharmacotherapy of Huntington's chorea.

Table 10.6. Receptor abnormalities in Huntington's chorea

Receptor	Striatal receptor binding	Author(s)
Cholinergic-muscarinic	↓ ↓	Hiley and Bird (1974) Wastek et al. (1976)
Serotonin	↓	Enna et al. (1976)
Beta-adrenergic	N	Enna et al. (1976)
GABA	↓ ↓ N	Lloyd et al. (1977) Olsen et al. (1979) Enna et al. (1977)
Dopamine	↓	Reisine et al. (1978)
Benzodiazepine	↓	Reisine et al. (1979)

N, normal; ↓, decreased.

At the present time, most knowledge is available concerning the effects of D2 receptor agonists and antagonists. Drugs such as bromocryptine which modify activity at the D2 receptor have been used in the treatment of neurological and endocrinological diseases. Parkinsonism and galactorrhoea are notable examples. The most commonly used drugs in Huntington's chorea, the phenothiazines, are both D1 and D2 receptor antagonists. If the neurological features of this disorder were primarily due to problems with neurotransmission at a specific dopamine receptor more successful treatment with lower frequency of unwanted side effects may result from the development of selective dopaminergic pharmacotherapy.

There is a decrease in both the affinity and density of benzodiazepine receptors in Huntington's chorea (Reisine et al. 1979). Benzodiazepines have been demonstrated to potentiate GABA effects on the central nervous system and this may be the underlying reason for the improvement in some patients' symptoms with benzodiazepine therapy. Further studies of the receptor and neurochemical alterations in Huntington's chorea will provide a more rational basis for the treatment of the disease.

10.7. Neuroendocrine Disturbances

The evidence for neurochemical disturbances in the nigrostriatal neurotransmitter systems in Huntington's chorea has been presented. The finding of pathological changes in the hypothalamus together with the knowledge that these dopaminergic and GABA-ergic pathways modulate the synthesis and release of different pituitary hormones has prompted numerous workers to examine anterior pituitary hormone regulation in Huntington's chorea.

Of all the neurotransmitters, dopamine plays a most important role in neuroendocrine control of pituitary hormone secretion. Dopamine stimulates growth hormone secretion, whilst prolactin release is generally accepted to be under tonic inhibitory control, also probably mediated via dopamine (Fig. 10.5). Recent studies suggest that GABA, on the other hand, exerts a stimulatory role in human prolactin regulation (Tamminga et al. 1979). Whether GABA operates via decreasing the synthesis and release of hypothalamic

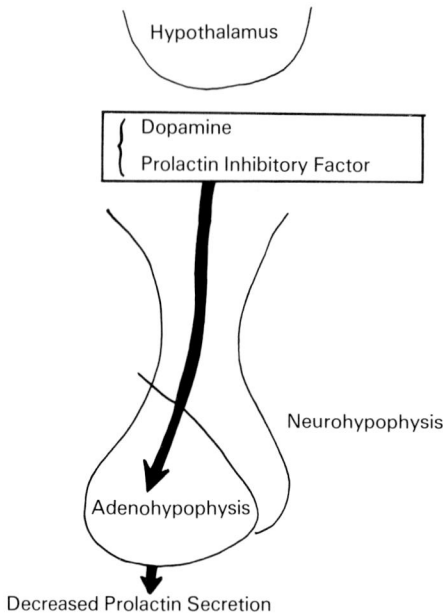

Fig.10.5. Dopamine inhibits prolactin secretion. (From Hayden and Vinik (1979) by courtesy of the Editors of *Advances in Neurology*, vol 23, and Raven Press)

dopamine or directly on the pituitary is uncertain. A review of the published reports of pituitary hormone function in Huntington's chorea is shown in Tables 10.7 and 10.8.

Whilst the results of these studies are not in complete agreement most investigations point to some disorder of neurotransmitter regulation of the hypothalamic-pituitary axis in this disease. Confounding factors may include prior neuroleptic therapy, the failure adequately to define clinical features, small patient samples, unsuitable selection of controls and normal individual variability in response.

Emotional and behavioural changes in Huntington's chorea are common and may herald the onset of the disease. Hormones exert influence on affect and conduct, quite apart from their classic endocrine function. Thus in some as yet unknown way disturbed hormonal release in Huntington's chorea may have a role in the evolution of the psychiatric symptomatology of the disease. Whilst these investigations have been most useful in assessing and evaluating neurotransmitter function and their interaction with the endocrines, and have provided some insight into the origin of motor and behavioural disorders, they have failed to substantially increase our knowledge and understanding of the primary defect in this disease.

10.8. Predictive Tests

One of the major goals of research endeavour in Huntington's chorea has been the search for a reliable method of determining whether a symptom-free descendant of an affected person carries the gene for the disorder. Presymptomatic detection of Huntington's chorea is based on the hypothesis that the abnormal gene exerts its effect long before the clinical features are manifest and that these aberrations are detectable. This, of course, may be untrue and the gene may lie in a dormant state until just prior to the onset of signs

Table 10.7. Review of published reports of prolactin regulation in Huntington's chorea

Prolactin responses in Huntington's chorea compared with normal subjects

Basal levels	TRH[a]	Chlorpro-mazine	Bromo-cryptine	L-Dopa	Apo-morphine	Placebo	Author(s)
↓	↓	↓					Hayden et al. (1977)
↑			↓				Caraceni et al. (1977)
N			N				Chalmers et al. (1978)
↑		↓	N	↓		↑	Caine et al. (1978)
↓							Paulson et al. (1979)
N			N				Leopold and Podolsky (1979)
N	N		N	N	N	N	Muller et al. (1979)
N			↓				Frattola et al. (1979)

N, the same as controls; ↑, greater than controls; ↓, less than controls.
[a] TRH, thyrotropin releasing factor.

Table 10.8. Review of published reports of growth hormone (GH) regulation in Huntington's chorea

GH responses in Huntington's chorea compared with normal subjects

Basal levels	Glucose	Insulin	L-Dopa	Bromo-cryptine	Apo-morphine	Chlor-promazine	TRH	Author(s)
N	↓		↑					Podolsky and Leopold (1974)
↑		↑						Phillipson and Bird (1976)
N		↑						Keogh et al. (1976)
N				↑				Caraceni et al. (1977)
N				↓				Chalmers et al. (1978)
N						↓	N	Hayden and Vinik (1979)
N			↑	↑	↑			Muller et al. (1979)

N, the same as controls; ↑, greater than controls; ↓, less than controls.

and symptoms of the disease. All attempts at devising a predictive test, however, presuppose the former hypothesis.

10.8.1. Ethical Dilemmas

The possible advent of a successful reliable predictive test poses numerous ethical problems. It could be argued that these issues are at present irrelevant as there is still no safe established method for presymptomatic identification of the heterozygote. However, without strict guidelines for the utilisation of such a test its implementation could have disastrous effects.

Advances in presymptomatic detection and treatment may not occur concurrently. In other words, it may be possible to identify the heterozygote without being able to offer any curative therapy. Whilst it would be most important to reassure those persons known to be unaffected, how would one counsel the affected? The hope of those in this group will be blighted at a time when they still have 15-20 healthy, creative years before them and may result in a lifetime of dread and anxiety or, possibly, even suicide. In other instances, however, such persons, informed of their heterozygote status, may make more responsible decisions and plan for their lives accordingly, on the basis of this accurate information.

Numerous questions arise which need to be considered. What would the impact of such a test be, were it positive, on social issues, such as marriage and reproductive behaviour? When and by whom should the test be done? What will the effect be on parents and siblings watching a known 'carrier' grow up? Will that knowledge alter the family's responses to that child's behaviour and moods? How would such a test influence economic issues, such as employment possibilities and life insurance? Would it result in discrimination against such persons in these areas? How would the knowledge that one has the abnormal gene influence that person's motivation in terms of further study and involvement in long-term projects? At the present stage all persons at risk for Huntington's chorea by virtue of their having an affected parent, have an even chance of inheriting the gene. These persons can all be encouraged to maintain some optimism that they may escape inheritance of the abnormal gene. It could be argued that in the current milieu, with no available curative therapy, this may be a more humane alternative.

Despite the dilemmas regarding utilisation of such a test it is essential that research in this area be continued, as it may constitute the most promising step towards understanding the primary defect of this disease and may facilitate development of more efficacious therapy. But such research is fraught with dangers which can and must be avoided. For this reason the Commission for the Control of Huntington's Disease (1977) has formulated preliminary broad guidelines for workers in this field:

1. All subjects must be entirely voluntary.

2. All potential subjects must understand there is at present no reliable valid predictive test for Huntington's chorea.

3. All proposed research must be submitted to an ethical and scientific review board prior to its implementation.

4. No information on results should be given to subjects, and volunteers should agree to this condition prior to participating in this research.

These guidelines protect volunteer subjects and also allow investigators to continue the search for a predictive test and the underlying biochemical aberration of Huntington's chorea.

A long-term goal of such research is development of a test for prenatal diagnosis and the subsequent option of termination of pregnancy. The awesome implication of such a screening procedure is that the at-risk parent is thus simultaneously diagnosed with the fetus.

Table 10.9. A review of the attempts at presymptomatic diagnosis of Huntington's chorea

Disturbance in affected persons	Technique used	No. of asymptomatic individuals showing some abnormality	Authors
Neurological			
Anomalies of conjugate ocular movement	Opticoelectronics	10/30	Petit and Milbled (1973)
Muscle tremor	Accelerometer	6/15	Myers and Falek (1979)
Abnormal muscle tone	EMG	2/19	Baro (1973)
Psychiatric/mental			
Personality change	Bender Gestalt, WISC, MMPI, ISB, DAP tests	5/12	Goodman et al. (1966)
	MMPI, Rorschach inkblot test	13/25	Palm (1973)
Dementia	Wechsler FSIQ	29/85	Lyle and Gottesman (1977)
Neurophysiological			
Excess slow low-voltage activity	EEG	19/26	Patterson et al. (1948)
	H. reflex	5/8	Johnson et al. (1977)
Neurochemical			
Decreased CSF GABA	Ion-exchange fluorometry	13/22	Manyam et al. (1977)
Neuroendocrinological			
Disturbed prolactin release	Immunoassay	11/23	Hayden et al. (1977)
Growth hormone hyper-responsiveness to bromo-cryptine	Immunoassay	4/13	Caraceni et al. (1979)
Neuropharmacological			
Aggravation of chorea	L.-Dopa administration	10/28	Klawans et al. (1972)
Neuroradiological			
Caudate nucleus atrophy	CAT scan	0/47	Neophytides et al. (1979)

10.8.2. Attempts at Presymptomatic Diagnosis

There have been numerous attempts at presymptomatic diagnosis of Huntington's chorea and these are reviewed in Table 10.9. In contrast to the plethora of publications describing these preliminary methods, there have been very few reports of follow-up studies of these initial tests (Table 10.10). A major factor which may account for this phenomenon is that most investigators have used children of affected persons and thus a long time must elapse prior to declaring the subjects gene-free.

The best-known and most controversial attempt at presymptomatic diagnosis is the L-dopa provocative test (Klawans et al. 1972), which was based on the rationale that

Table 10.10.   Follow-up results of predictive tests

Tests	No. tested	Abnormal result		Normal result		Author(s)
		No. found	No. affected	No. found	No. affected	
EEG	26	19	8	7	4	Chandler (1969)
L-Dopa	30	10	5	20	1	Klawans et al. (1980)
Eye movements	30	10	2	20	1	Petit (1979)
Psychological disturbance	85	29	15	56	12	Lyle and Gottesmann (1979)
	29	13	—	16	1	Palm (1973)
Neuroendocrine disturbance	23	11	1	12	1	Hayden (1979)

stimulation of receptors hypersensitive to dopamine might elicit chorea in those who have the abnormal gene. This test has fallen into disrepute for numerous reasons. Whilst the determination of false positive results necessitates long-term follow-up and is always difficult to prove, it has been clearly established that false negative results may occur with this test (Klawans et al. 1980). Another most severe drawback is that this method allows those at risk to read the results, which may impose severe psychological stress on these subjects. Furthermore the theoretical possibility that L-dopa may hasten onset of the disease is another reason for precluding this method as a predictive test.

Numerous attempts at presymptomatic diagnosis have been made but at the present time there is not a single completely reliable test which easily and unambiguously distinguishes between those persons with and those without the abnormal gene for Huntington's chorea.

10.9.   Conclusion

The words of Galileo, one of the fathers of the scientific method, could well be applied to the history of research endeavour in Huntington's chorea: "Yes, we will question everything, everything once again. And what we find today we shall strike out from the record tomorrow and only write it in again when we have once more discovered it" (Brecht: *Life of Galileo* 1938).

During the last few years there has been a dramatic change in the investigative approaches to this disorder. Whilst earlier studies focused attention on the pathology and the neurochemical correlates of the disease, recent advances in molecular biology have ushered in a new phase of research, namely that of cell biology and investigation of the extraneural changes in this disease. At the present time it is impossible to know where the secret to our understanding of the disorder lies and thus, whilst concentration of research effort in certain circumscribed promising areas is acceptable, a broad investigative approach is still warranted. If the current ferment of research activity into Huntington's chorea continues there is reason for optimism.

Great advances in the understanding of the pathophysiology of Huntington's chorea were originally made as a result of the discovery of dopamine deficiency in Parkinson's disease. Further elucidation of the underlying mechanisms of Huntington's chorea will in turn benefit research in other neurological and psychiatric diseases with shared features such as the presenile dementias, chorea due to other causes and schizophrenia.

Of the 2811 described genetic diseases, a total of 1489 disorders are inherited as autosomal dominant traits (McKusick 1978). In contrast to the tremendous advances in the understanding of autosomal recessive diseases, there is a dearth of knowledge concerning the basic defects in autosomal dominant conditions. Elucidation of the primary defect in Huntington's chorea would provide a conceptual framework for the investigation of other similarly inherited disorders.

The biochemical and neuroendocrine examination of patients with Huntington's chorea has enhanced our understanding of the interactions between different neurotransmitters and the endocrines. The functions of the neuropeptides and their interaction with the other neurotransmitters could also be further explored by the investigation of affected patients. In other words, Huntington's chorea can serve as a prototype for research and treatment of other psychiatric, neurological and genetic diseases and may provide greater insight into neuroendocrine function in health and disease (Fig. 10.6).

However, research is not the only area which binds Huntington's chorea to other disorders. The need for comprehensive care and community services is shared by all persons suffering from chronic, progressive diseases, and improvement of conditions for patients with Huntington's chorea is closely related to a greater awareness of the needs of all persons who suffer from such disorders.

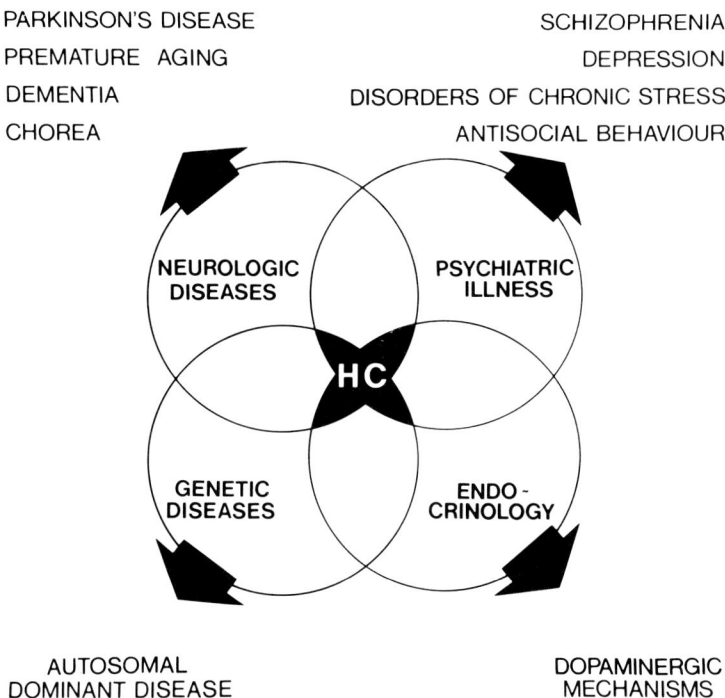

Fig. 10.6. Huntington's chorea as a prototype for the study and research into other genetic, neurological and psychiatric diseases

References

Anden NE (1974) Inhibition of the turnover of the brain dopamine after treatment with gamma-aminobutyrate: 2-oxyglutarate transamine inhibitor amino-oxyacetic acid. Arch Pharmacol 283: 419-424

Arlett CF, Muriel WJ (1979) Radiosensitivity in Huntington's chorea cell strains: a possible pre-clinical diagnosis. Heredity 42:276

Arlett CF (1980) Presymptomatic diagnosis of Huntington's disease? Lancet I: 540

Arregui A, Bennett JP, Bird ED, Tamumura HI, Iverson LL, Snyder SH (1977) Huntington's chorea: selective depletion of angiotensin convertng enzyme in the corpus striatum. Ann Neurol 2: 294-298

Arregui A, Iversen LL, Spokes EG, Emson PC (1979) Alterations in postmortem brain angiotensin-converting enzyme activity and some neuropeptides in Huntington's disease. In: Chase TN, Wexler NS, Barbeau A (eds) Advances in neurology, vol 23. Raven Press, New York, pp 517-526

Auerbach P (1980) Immunohistochemical study of foci of recent cell death in Huntington's disease. Canad J Neurol Sci 7(2): 87-88

Barkley DS, Hardiwidjaja S, Menkes JH (1977) Abnormalities in growth of skin fibroblasts of patients with Huntington's disease. Ann Neurol 1: 428-430

Barkley DS, Hardiwidjaja S, Tourtellotte W, Menkes JH (1978) Cellular immune responses in Huntington's disease. Neurology 28: 32-35

Baro F (1973) A neuropsychological approach to early detection of Huntington's chorea. In: Barbeau A, Chase TN, Paulson GW (eds) Huntington's chorea, 1872-1972. Raven Press, New York, pp 329-338

Bernheimer H, Hornykiewicz O (1973) Brain amines in Huntington's chorea. In: Barbeau A, Chase TN, Paulson GW (eds) Huntington's chorea, 1872-1972. Raven Press, New York, pp 525-553

Bernheimer H, Sperk G, Price KS, Hornykiewicz O (1979) Brain gangliosides in Huntington's disease. In: Chase TN, Wexler NS, Barbeau A (eds) Advances in neurology, vol 23. Raven Press, New York, pp 463-471

Beverstock G (1979) Communication to Eighth International Workshop on Huntington's Chorea, Oxford, England

Bird ED, Iversen LL (1974) Post mortem measurements of glutamic acid decarboxylase, choline acetyltransferase and dopamine in basal ganglia. Brain 97: 457-472

Bird ED, Iversen LL (1976) Neurochemical findings in Huntington's chorea. In: Youdim MBH, Lovenberg W, Sharman DF, Lagnado JR (eds) Essays in neurochemistry and neuropharmacology. Wiley, New York, pp 177-193

Bird ED, Caro AJ, Pilling JB (1973) A sex-related factor in the inheritance of Huntington's chorea. Ann Hum Genet 37: 255-259

Bird TD (1976) Normal glutamic acid decarboxylase activity in kidney tissue from patients with Huntington's disease. J Neurochem 27: 1555-1557

Brown WT, Ambruster J, Darlington GJ (1979) Two-dimensional analysis of radiolabeled proteins in cultured Huntington's disease fibroblasts. In: Chase TN, Wexler NS, Barbeau A (eds) Advances in neurology, vol 23. Raven Press, New York, pp 361-370

Butterfield DA, Oeswein JQ, Markesbery WR (1977) Electron spin resonance study of membrane protein alteration in erythrocytes in Huntington's disease. Nature 267: 453-455

Butterfield DA, Oeswein JQ, Prunty MJ, Hisle KC, Markesbery WR (1978a) Increased sodium plus potassium adenosine-triphosphatase activity in erythrocyte membranes in Huntington's disease. Ann Neurol 4: 60-62

Butterfield DA, Prunty MJ, Markesbery WR (1978b) Electron spin resonance haematological and deformability studies of erythrocytes from patients with Huntington's disease. Biochem Biophys Acta 551: 452-458

Caine E, Kartzinel R, Ebert M, Carter AC (1978) Neuroendocrine function in Huntington's disease: dopaminergic regulation of prolactin release. Life Sci 22: 911-918

Caraceni TA, Panerai AE, Parati EA, Cocchi D, Muller EE (1977) Altered growth hormone and prolactin responses to dopaminergic stimulation in Huntington's chorea. J Clin Endocrinol Metab 44: 870-875

Caraceni TA, Parati EA, Cocchi D, Mainini P, Muller EE (1979) Neuroendocrine correlates in Huntington's chorea. In: Muller EE, Agnoli A (eds) Neuroendocrine correlates in neurology and psychiatry. Elsevier, Amsterdam, pp 167-178

Cassiman JJ, Verlinden J, Vlietinck RF, Bellemans J, Van Leuven F, Deroover J, Baro F, Van Den Berghe H (1979). The qualitative and quantitative study of the growth and cell surface properties of Huntington's disease fibroblasts and age matched controls. Hum Genet 1: 75-86

Chalmers RJ, Johnson RH, Keogh HJ, Nanda RN (1978) Growth hormone and prolactin response to bromocryptine in patients with Huntington's chorea. J Neurol Neurosurg Psychiatry 41: 135-139

Chandler JH (1969) EEG in the prediction of Huntington's chorea. In: Barbeau A, Brunette JR (eds) Progress in neurogenetics, Vol I. Excerpta Medica, Amsterdam, pp 564-585

Comings DE (1979) A search for the mutant protein in Huntington's disease and schizophrenia. In: Chase TN, Wexler NS, Barbeau A (eds) Advances in neurology, vol 23. Raven Press, New York, pp 335-349

Crow TJ, Johnstone EC, Owens DG, Ferrier IN, MacMillan JF, Perry RP, Tyrell DA (1979) Lancet I: 842-844

Curtis DR (1976) Some aspects of the clinical neuropharmacology of amino acid neurotransmitters. In: Yahr MD (ed) The basal ganglia. Raven Press, New York, pp 163-183

Curzon G, Gumpert J, Sharpe D (1972) Amine metabolites in the cerebrospinal fluid in Huntington's chorea. J Neurol Neurosurg Psychiatry 35: 514-519

Diamond BI, Borison RL (1978) Enkephalins and nigrostriatal function. Neurology (Minneap) 28: 1085-1088

Diamond BI, Comaty JE, Sudakoff GS, Haudata HS, Walter R, Borison RL (1979) Role of substance P as a 'transducer' for dopamine in model choreas. In: Chase TN, Wexler N, Barbeau A (eds) Advances in neurology, vol 23. Raven Press, New York, pp 505-515

Emson PC, Rehfeld JF, Langevin H, Rossar R (1980) Reduction in cholecystokinin-like immunoreactivity in the basal ganglia in Huntington's disease. Brain Res 198: 497-500

Enna SJ, Bird ED, Bennett JP, Bylund DB, Yamamura HI, Iversen LL, Snyder SN (1976) Huntington's chorea: changes in neurotransmitter receptors in the brain. N Engl J Med 294: 1305-1309

Enna SJ, Stern LZ, Wastek GJ, Yamamura HI (1977) Cerebrospinal fluid and gamma-aminobutyric acid variations in neurological disorders. Arch Neurol 34: 683-685

Foerster K, Freudenberg J (1980) HLA antigen frequencies in patients with Huntington's chorea and their relatives. J Neurol 223: 119-125

Frattola L, Albizzati MG, Bassi S, Trabucchi M (1979) Plasma prolactin levels: control by dopamine and GABA in Huntington's chorea patients. In: Muller EG, Agnoli A (eds) Neuroendocrine correlates in neurology and psychiatry. Elsevier/North-Holland, Amsterdam, pp 159-162

Frederickson DS, Goldstein JL, Brown MS (1978) The familial hyperlipoproteinemias. In: Stanbury JB, Wyngaarden JB, Frederickson DS (eds) The metabolic basis of inherited disease, 4th edn. McGraw-Hill, New York, pp 604-655

Gajdusek C (1977) Report of the Commission for the Control of Huntington's Disease and its Consequences, vol II, Technical report. US Government Printing Office, Washington, DC, p 20

Glaeser BS, Vogel WH, Oleweiler DB, Hare TA (1975) GABA levels in cerebrospinal fluid of patients with Huntington's chorea: a preliminary report. Biochem Med 12: 380-385

Goetz J, Roberts E, Comings ED (1975) Fibroblasts in Huntington's disease. N Engl J Med 293: 1225-1227

Goetz I, Roberts E, Warren J, Comings DE (1979) Growth of Huntington's disease fibroblasts during their in vitro life span. In: Chase TN, Wexler N, Barbeau A (eds) Advances in neurology, vol 23. Raven Press, New York, pp 351-359

Goodman RM, Hall CL, Terango L, Perrine GA, Roberts PL (1966) Huntington's chorea: a multidisciplinary study of affected parents and first generation offspring. Arch Neurol 15: 345-355

Gray PN, Dana SL (1979) GABA synthesis by cultured fibroblasts obtained from persons with Huntington's disease. J Neurochem 33: 985-992

Harper RA, Grove G (1979) Human skin fibroblasts derived from papillary and reticular dermis: differences in growth potential in vitro. Science 204: 526-527

Hayden, MR (1979) Communication to Eighth International Workshop on Huntington's Chorea, Oxford, England

Hayden MR, Vinik AI (1979) Disturbances in hypothalamic-pituitary hormonal dopaminergic regulation in Huntington's disease. In: Chase TN, Wexler NS, Barbeau A (eds) Advances in neurology, vol 23. Raven Press, New York, pp 305-315

Hayden MR, Vinik AI, Paul M, Beighton P (1977) Impaired prolactin release in Huntington's chorea: evidence for dopaminergic excess. Lancet II: 423-426

Hiley CR, Bird ED (1974) Decreased muscarinic receptor concentration in post mortem brain in Huntington's chorea. Brain Res 80: 355-358

Husby G, Li L, Davis LE (1977) Antibodies to human caudate nucleus neurons in Huntington's chorea. J Clin Invest 59: 922-932

Husby G, Forre O, Williams RC (1979) IgG subclass variable H-chain subgroup and light chain-type composition of antineuronal antibody in Huntington's disease and Sydenham's chorea. Clin Immunol Immunopathol 14: 361-367

Igbal K, Tellez-Nagel I, Grundke-Igbal I (1974) Protein abnormalities in Huntington's chorea. Brain Res 76: 178-184

Johnson EW, Radecki PL, Paulson GW (1977) Huntington's disease: early identification by H-reflex testing. Arch Phys Med Rehabil 58: 162-166

Kanazawa I, Bird ED, O'Connell R, Powell D (1977) Evidence for a decrease in substance P content of substantia nigra in Huntington's chorea. Brain Res 120: 387-392

Kanazawa I, Bird ED, Gale JS, Iversen LL, Jessell TM, Muramoto O, Spokes EG, Sutoo D (1979) Substance P: decrease in substantia nigra and globus pallidus in Huntington's disease. In: Chase TN, Wexler NS, Barbeau A (eds) Advances in neurology, vol 23. Raven Press, New York, pp 495-504

Kebabian JW, Calne DB (1979) Multiple receptors for dopamine. Nature 277: 93-96

Keogh HJ, Johnson RH, Nanda RN, Sulaiman WR (1976) Altered growth hormone release in Huntington's chorea. J Neurol Neurosurg Psychiatry 39: 244-248

Kirk D, Parrington JM, Corney G, Bolt JMW (1977) Anomalous cellular proliferation in vitro associated with Huntington's disease. Hum Genet 30: 143-154

Kjellin KG, Stibler H (1974) CSF-protein patterns in extra-pyramidal diseases: preliminary report with special reference to the protein patterns in Huntington's chorea. Eur Neurol 12: 186-194

Klawans H, Goetz CG, Paulson GW, Barbeau A (1980) Levadopa and presymptomatic detection of Huntington's disease: eight-year follow up (letter) N Engl J Med 302: 1090

Klawans HL, Paulson GW, Ringel SP, Barbeau A (1972) Use of L-dopa in the detection of presymptomatic Huntington's chorea. N Engl J Med 286: 1332-1334

Leonardi A, De Martini IS, Perdelli F, Mancardi GL, Salvarani S, Bugiani O (1978) Skin fibroblasts in Huntington's disease. N Engl J Med 298: 632

Leopold NA, Podolsky S (1979) Levodopa and glucose influence on prolactin secretion in Huntington's disease. In: Chase TN, Wexler NS, Barbeau A (eds) Advances in neurology, vol 23. Raven Press, New York, pp 299-304

Lloyd KG, Dreksler ED, Bird ED (1977) Alterations in ^{3}H-GABA binding in Huntington's chorea. Life Sci 21: 747-754

Lloyd KG, Davidson L (1979) Alterations in ^{3}H-GABA binding in Huntington's disease: a phospholipid component? In: Chase TN, Wexler NS, Barbeau A (eds) Advances in neurology, vol 23. Raven Press, New York, pp 705-715

Lyle OE, Gottesman II (1977) Premorbid psychometric indicators of the gene for Huntington's disease. J Consult Clin Psychol 45: 1011-1022

Lyle OE, Gottesman II (1979) Subtle cognitive deficits as 15 to 20 year precursors of Huntington's disease. In: Chase TN, Wexler NS, Barbeau A (eds) Advances in neurology, vol 23. Raven Press, New York, pp 227-238

McGeer PL, McGeer EG, Fibiger HC (1973) Choline acetylase and glutamic acid decarboxylase in Huntington's chorea. Neurology (Minneap) 23: 912-917

McGeer PL, McGeer EG, Scherer U, Singh H (1977) A glutaminergic corticostriatal pathway? Brain Res 128: 369-373

McGeer EG, McGeer PL, Hattori T, Vincent SR (1979) Kainic acid neurotoxicity and Huntington's disease. In: Chase TN, Wexler NS, Barbeau A (eds) Advances in neurology, vol 23. Raven Press, New York, pp 577-591

McHenemey WH (1961) Immunity mechanisms in neurological disease. Proc R Soc Med 54: 127-136

McKusick VA (1978) Mendelian inheritance in man: catalogue of autosomal dominant, recessive and X-linked phenotypes, 5th edn. Johns Hopkins University Press, Baltimore

Magnusson T, Carllson A, Fisher GH, Chang D, Folkers K (1976) Effect of synthetic substance P on monoaminergic mechanisms in brain. J Neurol Transm 38: 89-93

Manyam NV, Hare TA, Katz L, Glaeser BS (1978) Huntington's disease: cerebrospinal fluid GABA levels in at-risk individuals. Arch Neurol 35: 728-730

Markesberry WR, Butterfield DA (1977) Scanning electron microscopy studies of erythrocytes in Huntington's disease. Biochem Biophys Res Commun 78: 560-564

Marsden CD (1979) Animal models of Huntington's disease: a review. In: Chase TN, Wexler NS, Barbeau A (eds) Advances in neurology, vol 23. Raven Press, New York, pp 567-576

Mattsson B (1974) Clinical, genetic and pharmacological studies in Huntington's chorea. UMEA University Medical Dissertations 7. UMEA, Sweden, pp 21-51

Menkes JH, Stein N (1973) Fibroblast cultures in Huntington's disease (letter). N Engl J Med 288: 856-857

Morrell RM (1979) Antibody-dependent cytoxicity in Huntington's disease. In: Chase TN, Wexler NS, Barbeau A (eds) Advances in neurology, vol 23. Raven Press, New York, pp 443-447

Moshell AN, Barrett SF, Tarone RE, Robbins JH (1980) Radiosensitivity in Huntington's disease: implications for pathogenesis and presymptomatic diagnosis. Lancet I: 9-11

Muller EE, Parati EA, Cocchi D, Zanardi P, Caraceni T (1979) Dopaminergic drugs on growth hormone and prolactin secretion in Huntington's disease. In: Chase TN, Wexler NS, Barbeau A (eds) Advances in neurology, vol 23. Raven Press, New York, pp 319-334

Myers RH, Falek A (1979) Quantification of muscle tremor of Huntington's disease patients and their offspring in an early detection study. Biol Psychiatry 14: 777-789

Neophytides AN, Di Chiro G, Barron SA, Chase TN (1979) Computer axial tomography in Huntington's disease and persons at-risk for Huntington's disease. In: Chase TN, Wexler NS, Barbeau A (eds) Advances in neurology, vol 23. Raven Press, New York, pp 185-191

Noronha AB, Roos RP, Antel JP, Arnason BG (1979) Huntington's disease: abnormality of lymphocyte capping. Ann Neurol 6: 447-450

Olney JW (1979) Excitotoxic amino acids and Huntington's disease. In: Chase TN, Wexler NS, Barbeau A (eds) Advances in neurology, vol 23. Raven Press, New York, pp 609-624

Olney JW, De Gubareff T (1978) The fate of synaptic receptors in the kainate-lesioned striatum. Brain Res 140: 340-343

Olsen RW, Van Ness P, Tourtellotte WW (1979) Gamma-aminobutyric acid receptor binding curves for human brain regions: comparison of Huntington's disease and normal. In Chase TN, Wexler NS, Barbeau A (eds) Advances in neurology, vol 23. Raven Press, New York, pp 697-704

Palm JD (1973) Longitudinal study of a preclinical test program for Huntington's chorea. In: Barbeau A, Chase TN, Paulson GW (eds) Huntington's chorea, 1872-1972. Raven Press, New York, pp 311-325

Patterson RM, Bagchi BK, Test A (1948) Prediction of Huntington's chorea: an electroencephalographic and genetic study. Am J Psychiatry 104: 786-797

Paulson GW (1979) Diagnosis of Huntington's disease. In: Chase TN, Wexler NS, Barbeau A (eds) Advances in neurology, vol 23. Raven Press, New York, pp 177-184

Pericak-Vance MA, Conneally PM, Merritt AD, Roos RP, Vance JM, Yu PL, Norton JA, Antel JP (1979) Genetic linkage in Huntington's disease. In: Chase TN, Wexler NS, Barbeau A (eds) Advances in neurology, vol 23. Raven Press, New York, pp 59-72

Perry T (1979) Communication to Eighth International Workshop on Huntington's Chorea, Oxford, England

Perry TL, Hansen S, Klosker M (1973) Huntington's chorea: deficiency of gamma-aminobutyric acid in brain. N Engl J Med 288: 337-342

Perry TL, Kish S, Hansen S, Buchanan J (1978) Brain gamma-aminobutyric acid deficiency in schizophrenia and Huntington's chorea. In: Proceedings of the Second International Symposium on Huntington's Disease, p A65

Petit H (1979) Communication to Eighth International Workshop on Huntington's chorea, Oxford, England

Petit H, Milbled G (1973) Anomalies of conjugate ocular movements in Huntington's chorea: application to early detection. In: Barbeau A, Chase TN, Paulson GW (eds) Huntington's chorea, 1872-1972. Raven Press, New York, pp 287-294

Pettegrew JW, Nichols JS, Stewart RM (1979) Fluorescence spectroscopy on Huntington's fibroblasts. J Neurochem 33: 905-911

Phillipson OT, Bird ED (1976) Plasma growth hormone concentration in Huntington's chorea. Clin Sci Mol Med 50: 551-554

Podolsky S, Leopold NA (1974) Growth hormone abnormalities in Huntington's chorea: effect of L-dopa administration. J Clin Endocrinol Metab 39: 36-39

Reisine TD, Fields JZ, Bird ED, Spokes E, Yamamura HI (1978) Characterisation of brain dopaminergic receptors in Huntington's disease. Comm Psychopharmacol 2: 79-84

Reisine TD, Beaumont K, Bird ED, Spokes E, Yamamura HI (1979) Huntington's disease: alterations in neuro-transmitter receptor binding in the human brain. In: Chase TN, Wexler NS, Barbeau A (eds) Advances in neurology, vol 23. Raven Press, New York, pp 717-726

Ringel SP, Weiner WJ, Rubovitz R, Klawans HL (1973) Methysergide in Huntington's chorea. In: Barbeau A, Chase TN, Paulson GW (eds) Huntington's chorea, 1872-1972. Raven Press, New York, pp 769-776

Roth RH, Ando N, Simon JR, Bird ED, Gold BI (1979) Gamma-hydroxybutyrate: alterations in endogenous brain levels in Huntington's disease. In: Chase TN, Wexler NS, Barbeau A (eds) Advances in neurology, vol 23. Raven Press, New York, pp 557-566

Sanberg PR, Fibiger HC (1979) Body weight, feeding and drinking behaviours in rats with kainic acid-induced lesions of striatal neurones. Exp Neurol 66: 444-446

Schneider EL (1979) Cell replication and aging: in vitro and in vivo studies. Fed Proc 38: 1857-1861

Spokes EG (1979) Dopamine in Huntington's disease: a study of postmortem brain tissue. In: Chase TN, Wexler NS, Barbeau A (eds) Advances in neurology, vol 23. Raven Press, New York, pp 481-493

Spokes EG (1980) Neurochemical alterations in Huntington's chorea: a study of postmortem brain tissue. Brain 103: 179-210

Tamminga C, Neophytides A, Chase TN, Frohman LA (1979) Stimulation of prolactin and growth hormone secretion by muscimol, a gamma-aminobutyric acid agonist. J Clin Endocrinol Metab 47: 1348-1351

Tourian A, Hung W (1977a) Glucosamine dependence of Huntington's chorea fibroblasts in culture. BBRC 76: 345-353

Tourian A, Hung W (1977b) Membrane abnormalities of Huntington's chorea fibroblasts in culture. BBRC 78: 1296-1303

Tourian A, Hung W (1979) Huntington's disease fibroblasts: nutritional and protein glycosylation studies. In: Chase TN, Wexler NS, Barbeau A (eds) Advances in neurology, vol 23. Raven Press, New York, pp 371-386

Valentine WN, Pagilia DC, Tartaglia AP, Gilsanz F (1977) Hereditary hemolytic anemia with increased red cell adenosine deaminase and decreased adenosine triphosphate. Science 195: 783-784

Wastek GJ, Stern LZ, Johnson PC, Yamamura HI (1976) Huntington's disease: regional alteration in muscarinic cholinergic receptor binding in human brain. Life Sci 19: 1033-1040

Weiner WJ, Hitri A, Carvey P, Koller WC, Nausieda PA, Klawans HL (1979) ^{3}H-dopamine binding studies in guinea pig striatal membrane suggesting two distinct dopamine receptor sites. In: Chase TN, Wexler NS, Barbeau A (eds) Advances in neurology, vol 23. Raven Press, New York, pp 687-695

Went LN, Volkers WS (1979) Genetic linkage. In: Chase TN, Wexler NS, Barbeau A (eds) Advances in neurology, vol 23. Raven Press, New York, pp 37-42

Wu JY, Bird ED, Chen MS, Huang WM (1979) Studies of neurotransmitter enzymes in Huntington's chorea. In: Chase TN, Wexler NS, Barbeau A (eds) Advances in neurology, vol 23. Raven Press, New York, pp 527-536

Appendix 1 The Use of Conditional Probabilities in Genetic Counselling for Huntington's Chorea

Let the at-risk person be a healthy 49-year-old man living in Northamptonshire, England, whose mother and grandfather were affected with Huntington's chorea. By the age of 49, 80% of all persons in that area who have the abnormal gene have already developed clinical signs of the disease (Table 4.2; Oliver 1970). Therefore he is either among the 20% of persons who have the gene for Huntington's chorea and who manifest later in life or he belongs to the normal population in the sense that he does not have the gene for this disorder. These factors can be included in estimating his remaining risk of developing the disease.

His *prior probability* of having the abnormal gene = 0.5 (50%)

The *conditional probability* of him being asymptomatic if heterozygous = 0.20 (20%)

The total *posterior probability* of him being asymptomatic and not having the abnormal

gene $= \dfrac{0.5 \times 0.2}{0.5 \times 0.2 + 0.5 \times 1.0}$

$= 0.16 \ (16\%)$

He therefore has a 16% chance of still developing Huntington's chorea, whilst the risk estimate for his children is thus 8%.

Another way of calculating this figure is via a diagram, such as that shown in Fig. A.1 (Vogel and Motulsky 1979). The large square is divided into 100 small squares each of which represents 1%. At birth this man has a 50% risk of developing Huntington's chorea at some stage in his life, represented by the 50 lightly shaded small squares in the left-hand square (A). At the age of 49 he has survived 80% of the potential manifestation period without becoming affected. This is represented by the dark shading in the right-hand square (B).

His chances of being unaffected but still heterozygous for Huntington's chorea are portrayed by the 10 lightly shaded squares in B. The possibility that he does not have the abnormal gene is represented by the 50 unshaded squares in A and B.

His total probability of being asymptomatic yet heterozygous for the disorder is

$$\frac{10 \ \text{squares in the lightly shaded area}}{50 \ \text{squares unshaded} + 10 \ \text{squares in the lightly shaded area}}$$

$= 0.16 = 16\%.$

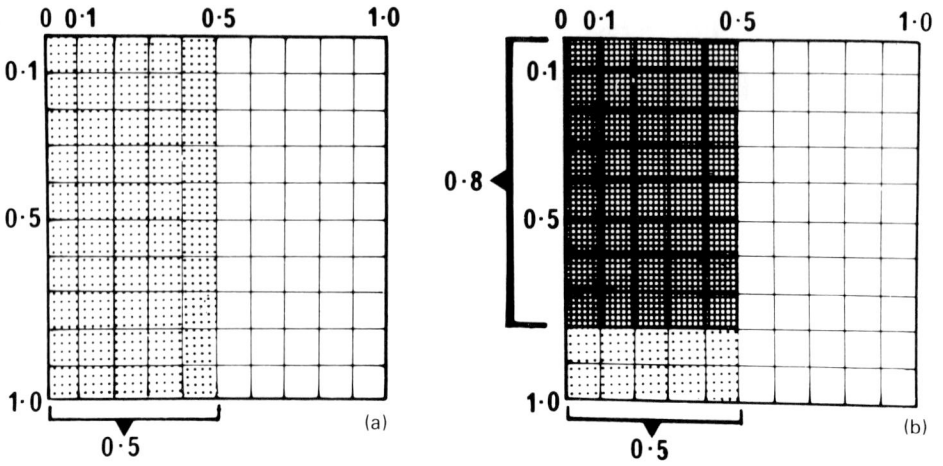

Fig.A.1. Diagram showing the principle of risk calculation for children of a parent with Huntington's chorea

To be strictly accurate, a knowledge of the ranges of ages at onset of Huntington's chorea in the particular area under study is required in order to make these calculations. However such investigations have only been conducted in a few regions, as seen in Table 4.2, p.49). The average figures of the different investigations listed in the table are useful as approximations where such details are unavailable.

Reference

Vogel F, Motulsky A (1979) In: Human genetics: problems and approaches. Springer, Berlin Heidelberg New York, pp 595-596

Apppendix 2 Method for Determination of the Mutation Rate in Huntington's Chorea

Previous authors (Mattsson 1974; Reed and Neel 1959; Stevens 1976) have used two different methods for estimation of the mutation rate. The easiest and most direct way makes use of the formula

$$Mu = \frac{P \times F}{2},$$

where Mu is the mutation rate, P is the proportion of affected individuals in the survey who did not have affected parents, and F is the heterozygote frequency.

The second method, which presumes that the population is in equilibrium, depends on the assessment of biological fitness of patients with Huntington's chorea in the given community.

$$Mu = \frac{(1 - W) F}{2},$$

where W is the fitness of patients with chorea compared with the fitness of the general population, and F is the heterozygote frequency.

References

Mattsson B (1974) Clinical, genetic and pharmacological studies in Huntington's chorea. UMEA Medical Dissertations 7. UMEA, Sweden, pp 21-51

Reed TE, Neel JV (1959) Huntington's chorea in Michigan. II. Selection and mutation. Am J Hum Genet 11: 107-136

Stevens DL (1976) Huntington's chorea: a demographic, genetic and clinical study. MD thesis, University of London, pp 1-338

Appendix 3   Method for Determination of the Heterozygote Frequency in Huntington's Chorea

Whilst different methods have been used to calculate the heterozygote frequency in Huntington's chorea, the most simple and efficient is that initially described by Reed et al. (1958) as follows:

$$F = \frac{H}{x\ Nx\ Px},$$

where F is the frequency of heterozygotes in the general population, H is the number of observed patients with Huntington's chorea in a given area, Nx is the total number of individuals who are age x, and Px the proportion of heterozygotes whose chorea is recognised by age x.

A major disadvantage of this technique is that its accuracy is directly related to the completeness of patient ascertainment, and underestimation of the total number of affected persons will inevitably result in artificial lowering of the heterozygote frequency.

Reference

Reed TE, Chandler JH, Hughes EM, Davidson RT (1958) Huntington's chorea in Michigan: demography and genetics. Am J Hum Genet 10: 201-225

Appendix 4 Names and Addresses of Lay Organisations and Other Centres for Information on Huntington's Chorea

United States

Committee to Combat Huntington's Disease, 250 West 57th Street, New York, NY 10019
Tel: 212-757 0443

National Huntington's Disease Association, 1441 Broadway Suite 501, New York, NY 10018
Tel: 216-966 4320

Wills Foundation,
PO Box 66704, Houston, Texas 77006
Tel: 713-965 9043

Hereditary Disease Foundation,
9701 Wilshire Boulevard, Suite 1204, Beverley Hills, California 90212
Tel: 213-274 5443

There are over 30 chapters and representatives of the Committee to Combat Huntington's disease (HD) throughout the United States. For further information in this regard contact the National Office in New York.

Canada

Huntington Society of Canada,
13 Water Street North, PO Box 333, Cambridge (Galt) Ontario, N1R 5T8
Tel: 519-6221002

United Kingdom

The Association to Combat Huntington's Chorea, 'Borough House', 34a Station Road, Hinckley, Leicestershire LE10 1AP
Tel: 0455 615558

Combat Holiday Home, Mrs. M. Jackson (Manageress), Theydon Road, Epping, Essex CM16 4DX
Tel: (0378)77588

The Huntington's Chorea Society,
Mr. G. Harris (Chairman), 12 Madeley Road, High Acres, Kingswinford, West Midlands

South East of Scotland Huntington's Chorea Association, Miss Elizabeth Findlay (Secretary), 17 Darnell Road, Edinburgh EH5 3PQ
Tel: 031-552 6885

Northern Ireland: Mrs. P. Page,
7 Craigdarragh Park, Glencraig, Holywood, Co. Down
Tel: (02317) 2150

Wales: Mrs. J. Barnes, The Rectory Church Street, Glyncorrwg, Port Talbot, W. Glamorgan
Tel: (063)983 423

There are numerous branches and area representatives of the Association to Combat Huntington's Chorea in the United Kingdom. For more information concerning the representatives in your area, contact the National Office in Middlesex.

Netherlands

Vereniging van Huntington, PA
Verkuyllaan 30, 1171 EE Badhoevedorp
Tel: 02960-2921

Secretary General:
Antonie Heinsuisstraat 6,
1052 EP Amsterdam
Tel: 020-826786

Belgium

Huntington Liga, c/o Psychiatrisch
Centrum, St. Kamillus, Krijkelberg 1,
B-3043 Bierbeek
Tel: 237847

West Germany

c/o Prof. H. Oepen, Bahnhofstrasse 7a,
355 Marburg

France

Association Huntington de France,
13 rue du 11 Novembre, 59650
Ville Neuve d'Ascq

Italy

Associazione per Combattere la Corea
Di Huntington, Presso Instituto
Neurologico 'Besta', Via Celoria 11,
20133 Milano
Tel: 2360341

South Africa

Department of Human Genetics,
University of Cape Town, Medical
School, Observatory, 7925, Cape Town
Tel: 472350

Australia

Australian Huntington's Disease
Association, Mary's Mount.
25 Yarrbat Avenue, Balwyn, Victoria
Tel: Business 341 56 63, 341 55 10,
341 55 11; Home 232 84 15

AHDA, Box 2648X, GPO Melbourne,
3001

Regional Organizations

New South Wales: Secretary, Australian
Huntington's Disease Association,
78 Damien Avenue,
Wentworthville 2145

Queensland: Secretary Australian
Huntington's Disease Association,
48 Barkala Street, The Gap, 4061

South Australia: Secretary Australian
Huntington's Disease Association,
46 Elizabeth Street, Evandale, 5069

Western Australia: Secretary Australian
Huntington's Disease Association,
35 Roseberry Street, Jolimont, 6014

Tasmania: Secretary Australian
Huntington's Disease Association,
PO Box 496, Launceston, 7250

New Zealand

New Zealand Huntington's Disease
Association, 23 Konene Street,
Rotorua

Appendix 5   Brain Donation Programme

Any questions or problems concerning the brain donation programme for Huntington's chorea should be directed to:

United States
The Brain Tissue Bank, Dr. Edward D. Bird, McLean Hospital, Belmont, Massachusetts 02178
Tel: (617)855-2400

There is 24-hour telephone coverage and collect calls will be accepted

United Kingdom
The MRC Brain Tissue Bank, Dept of Neurological Surgery and Neurology, Addenbrooke's Hospital, Hills Road, Cambridge CB2 2QQ
Tel: Cambridge 40011

Subject Index